# HANDBOOK OF PRACTICAL CB SERVICE

# HANDBOOK OF PRACTICAL CB SERVICE

**JOHN D. LENK**

*Consulting Technical Writer*

Prentice-Hall, Inc., Englewood Cliffs, N.J.

*Library of Congress Cataloging in Publication Data*

LENK, JOHN D
    Handbook of practical CB service.

    Includes index.
    1. Citizens band radio—Equipment and supplies—
Maintenance and repair.  I. Title.
TK6570.C5L46      621.3845′4′028      77-21292
ISBN  0-13-380568-9

© 1978 by Prentice-Hall, Inc., Englewood Cliffs, N.J. 07632

All rights reserved. No part of this book
may be reproduced in any form or
by any means without permission in writing
from the publisher.

Printed in the United States of America

10  9  8  7  6  5  4  3  2  1

PRENTICE-HALL INTERNATIONAL, INC., *London*
PRENTICE-HALL OF AUSTRALIA PTY. LIMITED, *Sydney*
PRENTICE-HALL OF CANADA, LTD., *Toronto*
PRENTICE-HALL OF INDIA PRIVATE LIMITED, *New Delhi*
PRENTICE-HALL OF JAPAN, INC., *Tokyo*
PRENTICE-HALL OF SOUTHEAST ASIA PTE. LTD., *Singapore*
WHITEHALL BOOKS LIMITED, *Wellington, New Zealand*

To Irene, the Sandpiper Lady

# CONTENTS

**PREFACE**

**CHAPTER 1   INTRODUCTION TO CB SERVICE   1**

| | |
|---|---|
| 1-1 Rules and definitions | 2 |
| 1-2 Operation of a typical CB set | 8 |
| 1-3 The basic troubleshooting functions | 16 |
| 1-4 The universal troubleshooting procedure | 18 |
| 1-5 Applying the troubleshooting approach to CB service | 24 |
| 1-6 Trouble symptoms | 24 |
| 1-7 Localizing trouble | 27 |
| 1-8 Isolating trouble to a circuit | 35 |
| 1-9 Locating a specific trouble | 46 |

**CHAPTER 2   CB SERVICE EQUIPMENT   65**

| | |
|---|---|
| 2-1 Safety precautions in CB service | 66 |
| 2-2 Signal generators | 68 |
| 2-3 Oscilloscopes | 72 |
| 2-4 Meters | 80 |
| 2-5 Probes | 82 |

| | |
|---|---:|
| 2-6 Frequency meters and counters | 90 |
| 2-7 Dummy load | 95 |
| 2-8 RF wattmeter | 97 |
| 2-9 Vacuum tube, transistor, capacitor, and crystal testers | 98 |
| 2-10 Field strength meter | 100 |
| 2-11 Standing wave ratio (SWR) measurement | 101 |
| 2-12 Dip meters | 104 |
| 2-13 Miscellaneous CB service equipment | 107 |
| 2-14 Special test sets for CB service | 109 |

### CHAPTER 3   TYPICAL CB CIRCUITS   122

| | |
|---|---:|
| 3-1 Vacuum-tube CB circuits | 122 |
| 3-2 AM solid-state mobile CB circuits | 134 |
| 3-3 AM solid-state base station CB circuits | 147 |
| 3-4 Phase locked loop (PLL) CB circuits | 158 |
| 3-5 Single sideband (SSB) CB circuits | 168 |
| 3-6 Walkie-talkie CB circuits | 186 |

### CHAPTER 4   BASIC CB SERVICE APPROACH   192

| | |
|---|---:|
| 4-1 Servicing notes | 193 |
| 4-2 Frequency synthesizer, oscillator, and PLL troubleshooting | 215 |
| 4-3 Audio/modulation section troubleshooting | 219 |
| 4-4 Receiver section troubleshooting | 244 |
| 4-5 Transmitter section troubleshooting | 253 |

### CHAPTER 5   SERVICING ANTENNA AND NOISE PROBLEMS   271

| | |
|---|---:|
| 5-1 Antenna service problems | 271 |
| 5-2 Noise and interference problems | 276 |

### CHAPTER 6   CB SERVICE LITERATURE AND DATA   289

| | |
|---|---:|
| 6-1 CB channel frequencies | 289 |
| 6-2 Interpreting CB service literature | 290 |
| 6-3 Sample alignment/adjustment procedures | 294 |

# PREFACE

The purpose of this book is to provide a simplified system of service for many types and models of Citizen's Band (CB) radio sets in use. Originally, all CB sets used vacuum tubes and provided only amplitude-modulation (AM). Later, solid-state sets were added, and single sideband (SSB) communication was provided in some sets. Today, many CB sets include a number of special features such as digital channel selection and phase-locked loop (PLL) frequency control.

It is not only impractical, it is virtually impossible, in one book, to cover service for all such sets and their related circuits, although such attempts have been made. Rapid technological advances have soon made those books obsolete.

To overcome this problem, we concentrate on a *basic approach to CB service;* an approach that can be applied to any set: those now being manufactured, those to be manufactured in the future, and sets now in use. Keep in mind that older vacuum-tube sets, even though now manufactured in limited quantities, still require service (perhaps more so than the latest solid-state models).

The service approach described here is based on the techniques found in the author's highly successful troubleshooting books: *Handbook of Practical Solid State Troubleshooting, Handbook of Basic Electronic Troubleshooting,* and *Handbook of Simplified Television Service.*

As is described in Chapter 1 of this book, the author's techniques follow a *basic troubleshooting sequence*, which includes *failure symptom analysis, localizing* trouble to a section in the set (receiver, transmitter, audio/modulation, etc.), *isolating* trouble to a circuit within the section, and *locating* the specific trouble within the circuit.

Chapter 2 of the book is devoted to test equipment used in CB service. This is particularly important since CB faults are often best located by analyzing test results (response to input signals, voltage and resistance measurements, modulation measurements, etc.). Chapter 2 includes a wide variety of test equipment for practical CB work, and covers basic operating principles and characteristics. The discussions in Chapter 2 describe how the features found on present-day test equipment relate to specific problems in CB service.

Chapter 3 describes the theory of operation for a number of CB sets. A cross-section of present-day CB circuits is covered, including AM, SSB, PLL, mobile, base station, and walkie-talkie circuits.

Chapter 4 describes the basic approach to CB service, or how the basic troubleshooting procedures of Chapter 1 are combined with the practical use of test equipment discussed in Chapter 2, together with a knowledge of typical CB circuits covered in Chapter 3, to locate specific faults in various types of CB sets. Chapter 4 also describes alignment, adjustment, and test procedures required for CB service. Considerable emphasis is placed on "universal" tests, service and troubleshooting procedures that apply to all CB sets.

Chapter 5 describes service for CB antennas, and discusses electrical noise problems (particularly noise or interference caused by automobile engines). Specific, practical cures are given for electrical noise problems.

Chapter 6 is devoted to understanding and using CB service literature. The service procedures for a cross-section of CB sets is given. Using these examples, the reader should be able to relate the procedures to a similar set of circuits and controls on most CB sets (both vacuum-tube and solid-state).

Many professionals have contributed their talent and knowledge to the preparation of this handbook. The author acknowledges that the tremendous effort to make this book such a comprehensive work is impossible for one person, and he wishes to thank all who have contributed directly and indirectly.

The author wishes to give special thanks to the following: Apelco, Browning Laboratories, Courier, Craig Corp., Dynascan Corp., E. F. Johnson Co., Estes Engineering Co., Fieldmaster, Handic of U. S. A., Heath Company, Hy-Gain Electronics Corp., J. I. L. Corp., Kris, Inc., Lafayette Radio Electronics Corp., Midland International Corp., Olson Electronics, Pace Communications, PAL Electronics, Panasonic,

PREFACE xi

Pearce-Simpson, Radio Shack, Ray Jefferson, Regency Electronics, Inc., Robyn International, Inc., Royce Electronics Corp., SBE Linear Systems, Inc., Shakespeare Industrial Fiberglass Division, Sharp Electronics Corp., Sonar Radio Corp., Sony Corporation of America, Standard Communications, Teaberry Electronics Corp., Tram/Diamond Corp., and XTAL. The author also wishes to thank Mr. Joseph A. Labok of Los Angeles Valley College.

JOHN D. LENK

# 1
# INTRODUCTION TO CB SERVICE

This chapter is devoted to the basics of CB service. On the assumption that you may not know the rules regarding CB (frequencies used, power output allowed, who may service what, etc.), we will start with a summary of these rules. We will not go into any detail here since you must have copies of the full FCC regulations to operate or service CB equipment.

Next, we will describe operation of a typical CB set. The description is kept to the block-diagram level, since you are interested in service (not theory), and so that the descriptions apply to the greatest number of CB sets (new and old, vacuum-tube, solid-state, amplitude modulation, single sideband, 23- or 40-channel, etc.), CB circuits are covered in much greater detail throughout Chapter 3.

It is not intended that this introduction provide a complete course in electronic communications (broadcast and reception), since you must have a First-class or Second-class Radiotelephone Operator's license to service CB. If you have passed the technical examination for either of these licenses, you must know communications! However, you may not know how a typical CB set operates.

With CB basics established, we will describe the author's universal troubleshooting approach in detail. This is followed by discussions of how the universal troubleshooting approach can be applied to the specifics of CB service. Throughout these discussions, references are made to the remaining chapters where you will find detailed information on troubleshooting and service procedures.

## 1-1 RULES AND DEFINITIONS

This handbook covers equipment licensed as a Class-D Citizens Radio station under Part 95 of the FCC Rules and Regulations. Most present-day CB equipment falls into this class, although there are other classes. For example, there is Class C, which covers radio-control transmitters, Class A, which operates on frequencies in the 460 to 470 MHz band, unlicensed stations operating in the 160 to 190 kHz band, and so on.

### 1-1.1 Class-D Equipment and License

Class-D CB equipment operates in the 27 MHz band (the exact frequencies are given in Chapter 6) and is used for two-way radiotelephone communications. Typically, a CB set is a radio *transceiver* (combination transmitter and receiver) housed in a single case as shown in Fig. 1-1.

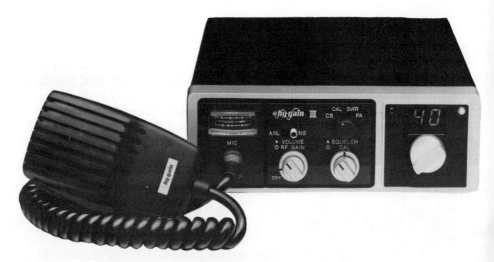

**Figure 1-1:** Hy-Gain III 40-channel Citizens Band mobile transceiver

The typical CB set operates on one or more of 23 channels (pre-1977) or 40 channels (those manufactured after January of 1977). Transmission and reception take place on the same channel, but not simultaneously. Thus, CB communications are classed as simplex. (When transmission and reception are simultaneous, this is called duplex operations.) Using

Sec. 1-1  Rules and Definitions    3

a simplex system, it is necessary to press a button (usually the microphone button) to transmit and release the button to receive. Class-D equipment requires a *station* license, but not an *operator's* license.

### 1-1.2 Walkie-talkie (Hand-held) Equipment

There are two basic types of walkie-talkie CB equipment. One type is the low power (100 mW or less input to the transmitter), with an antenna of five feet or less including the transmission line. These sets are covered by Part 15 of the FCC Rules and Regulations, and do not require either a station license or an operator license. However, such unlicensed sets can not be used (legally) for communications with a licensed CB (either transceiver or walkie-talkie). The other type of walkie-talkie meets the same general technical requirements as the typical transceivers and is covered by Part 95. It uses higher power (typically 1 or 2 W, but possibly up to 5 W), must be covered by a station license, and can be used to communicate with other licensed CB stations. Both types of walkie-talkies use the 27 MHz band.

### 1-1.3 Class-D Modulation and Power

Class-D CB equipment uses amplitude modulation (AM), defined as A3 by the FCC, with a power output limited to 4 W. Many Class-D CB sets also use single sideband (SSB), defined as A3a, with a power output limited to 12 W peak envelope power (PEP). The maximum permitted modulation frequency is 3 kHz, which covers voice use in normal conversations.

### 1-1.4 Class-D Tuning and Frequency Control

All Class-D transceivers are crystal controlled, and most are fixed tuned to a specific channel. That is, when a particular channel is selected, *both* the transmitter and receiver are tuned to the appropriate frequency and locked onto that frequency by crystals. Transmitter frequency stability must be 0.005% or better on all channels.

Some older CB sets have continuous-tuning or variable tuning where it is possible to tune the receiver over the entire frequency range (all channels). This has generally been replaced by fixed tuning. However, some CB sets provide for fine tuning of the receiver on either side of each channel (similar to a TV receiver). Other CB sets provide for a so-called Delta-tune control, which sets the receiver to some frequency on either side of the channel (to bring in transmitter signals that are slightly off frequency).

### 1-1.5 Types of Stations

The license for Class-D stations considers all transceivers as *mobile* units. This applies whether the set is on a boat, in an automobile, carried from place to place by hand, or is used at some fixed location. For our purposes, there are four types of stations in common use. A *mobile* station is one that can be transported readily and is either temporarily or permanently installed in a vehicle. The transceiver in Fig. 1-1 is a true mobile unit. Limited service can be performed on an installed mobile unit, but extensive service requires that the unit be removed from the auto or boat. A true *portable* station is hand-held, such as the walkie-talkie shown in Fig. 1-2, and is not designed for either temporary or permanent installation. Typically, the portable unit is brought into the shop for service.

**Figure 1-2:** Pace CB-155 portable (hand-held walkie-talkie) CB transceiver

A *fixed* station is used at one location only, and is intended for communications to other fixed stations. If the same unit is used for communication to other fixed stations, as well as mobile and portable

Sec. 1-1   Rules and Definitions                                                          5

stations, the unit is then called a *base* station. The transceiver shown in Fig. 1-3 is a typical base station or fixed station, depending on the use. Base or fixed stations can be serviced at the operating location, but extensive service is best done in the shop.

**Figure 1-3:**   Cobra 89XLR and 86XLR fixed-station (base-station) CB transceivers

Generally, a base or fixed station unit is connected to an outside antenna. However, some units are provided with an indoor antenna for short-range (typically 1 mile or less) operation. Mobile stations are connected to an antenna on the vehicle. Portable stations have built-in antennas (usually the telescope type).

Most base or fixed-station CB sets operate from 120 V, 60 Hz house power. Most mobile sets operate on 12 V dc power supplied by the vehicle battery and/or engine electrical system. (Note that the term 12 V is used for convenience. Most CB sets will operate on voltages in the 12-to-14 V range, and 13.8 V is typical.) Portable sets operate on internal batteries. There are a number of power packs (inverters, converters, etc.) that convert 120 V ac power into 12 V dc power, and vice versa. This permits mobile sets to be used as base or fixed stations, and vice versa. In most cases, the power packs are sealed units and can not be serviced. You simply replace the entire unit. From a practical standpoint, even if you can service the power pack, you probably can not find the replacement parts. This condition is true for many CB accessories (and even some CB sets)!

### 1-1.6 Type Acceptance

All Class-D transmitters sold or licensed after November 1974 must be *type accepted* by the FCC. Any transmitter not type accepted must not be operated after November 1978.

We will not discuss type acceptance in any detail here, except as it applies to service. In brief, the manufacturer (or an independent testing laboratory acting in behalf of the manufacturer) makes comprehensive tests of a transmitter (frequency stability, modulation, etc.) and submits the full details to the FCC. If the FCC approves, a type acceptance number is assigned to the transmitter design.

From a service standpoint, once a type acceptance number has been assigned, any change in the design not made with FCC permission, or any modifications made to the equipment, will void the type acceptance and make operation of such a transmitter illegal. In brief, if you replace any part with another part that is not an exact duplicate of the original, or a part not recommended by the manufacturer, type acceptance can be voided, and operation of the transmitter is illegal.

Obviously, type acceptance can raise some service problems. For example, some older CB sets may no longer be manufactured, and spare parts may no longer be available. (Many CB manufacturers are no longer in the business!) The RF coils and transformers (and other parts that affect frequency stability, modulation, etc.) are a particular problem. Of course, if you have full technical details of a part (such as winding data, impedances, etc.), it is possible to duplicate the part, but you must take full responsibility for replacement of the part. It is possible to get full technical details for most CB sets (even the old ones long out of manufacture) by means of Sams Photofacts, available from Howard W. Sams & Co., Inc., 4300 West 62nd St., Indianapolis, Indiana 46206. The author highly recommends that you get a copy of the appropriate Photofact service manual as a first step in servicing any CB set, not merely to solve a type acceptance problem.

Another problem may arise if you are called upon to modify or test a pre-1974 CB set to get type acceptance. The author does not recommend accepting such a job unless you have had extensive experience in communications service and laboratory testing. The same situation may arise if you attempt to modify a type-accepted 23-channel set for operation on 40 channels. Any modification of a type-accepted set automatically voids the original type acceptance. A possible exception may occur if the manufacturer provides a 40-channel modification kit or unit that has been type accepted. If you really want to become involved with type acceptance work, the author recommends a copy of the FCC Type Acceptance Manual, available from CB Magazine, 15 S. Overmyer Drive, Oklahoma City, Oklahoma 73127.

Sec. 1-1    Rules and Definitions

### 1-1.7    Who Can Service What

A First-class or Second-class Radiotelephone Operator's license is required to make any repairs or adjustments to a CB transmitter. The work may be done by an unlicensed person, but he must be supervised by a licensed operator. Also, the licensed operator must take full responsibility for the work. For example, if a CB transmitter is cited by the FCC for off-frequency operation, over-modulation, etc., the licensed operator who supervised the repair or adjustment must take the rap!

There are certain service procedures which may be made by anyone, without supervision by a licensed operator. The following are some examples:

Anyone may measure transmitter frequency or modulation level (transmitter frequency must be within 0.005% or better; modulation must not exceed 100%). A license is required to make any adjustments that affect frequency or modulation level. A possible exception to this rule occurs if modulation level can be set by a front panel control. In general, any front panel or external control is considered an operating control, and it can be set or adjusted by anyone. Any internal control must be set or adjusted by a licensed operator. There are exceptions of course, but this is a good rule of thumb.

Anyone may replace a vibrator (older CB sets often use vibrators in their power supply). Replacement of a plug-in vacuum tube or transistor may require a license. In some cases, replacement of a tube or transistor in a transmitter circuit can affect frequency or modulation level and will require adjustment of the transmitter circuits.

Anyone may replace *certified* crystals. The accuracy of such crystals is certified by the crystal manufacturer, and the CB manufacturer certifies that the crystals will operate properly in their particular transceiver. *Uncertified* crystals may be just as accurate and work just as well, but they will require test and possible adjustment after installation. Thus, uncertified crystals should be installed only by licensed operators.

Anyone may install or service the antenna system. Likewise, anyone may make the necessary tests after the antenna system is installed (to measure standing wave ratio, power output, proper transmission and reception, etc.).

Anyone may service the receiver circuits of a CB transceiver. However, this brings up a problem, since most modern CB units have circuits that are common to both the receiver and transmitter sections. Crystal control (frequency synthesizer) circuits are a typical example. Any service (parts replacement or adjustment) on these circuits will affect both the transmitter and receiver operation and requires a licensed operator.

### 1-1.8 Linear Amplifiers and Modulation Boosters

When used in connection with CB service, a *linear amplifier* is any broadband, RF amplifier (usually high power) used to increase the output of the CB transceiver. Typically, the linear amplifier is connected between the CB transceiver and the antenna, and delivers several hundred watts of power to the antenna. A linear amplifier will pass both the carrier and the modulation sidebands, and it thus extends the operating range of the CB unit. *Linear amplifiers are absolutely illegal for any type of CB operation!* The mere possession of a linear amplifier in connection with CB equipment is evidence of its use, and is thus illegal. The author recommends that you do not service linear amplifiers, or even accept them in your CB service shop.

A *modulation booster* is any device used to increase voice modulation of the CB transmitter. Typically, a modulation booster is an amplifier connected between the microphone and the CB transmitter modulation circuits. Ideally, a modulation booster will increase weak voice signals (caused by holding the microphone too far from the mouth) so as to produce the full 100% modulation. A well-designed modulation booster will also contain limiter circuits that prevent overmodulation. Operation of modulation boosters with CB sets is not illegal, but such operation can result in a problem if it causes overmodulation. If you service or install any type of modulation booster, always check the percentage of modulation (preferably using a loud voice close to the microphone). If the modulation is no more than 100%, you have no problem in regard to FCC rules. (If modulation is considerably less than 100% under these test conditions, you have a very poor modulation booster!)

### 1-2. OPERATION OF A TYPICAL CB SET

Figure 1-4 is the block diagram of a typical CB set shown in Fig. 1-5. This particular set is designed as a base or fixed station and uses vacuum tubes rather than solid-state components. The present trend is toward all solid-state CB. However, there are millions of vacuum-tube CB sets now in use, and all of them require service (perhaps more service than solid-state sets). This particular example is chosen since the circuitry is relatively simple and quite straight-forward.

#### 1-2.1 General Description

The set shown in Figs. 1-4 and 1-5 is a 23-channel crystal-controlled transmitter and receiver unit (transceiver) with a self-contained power

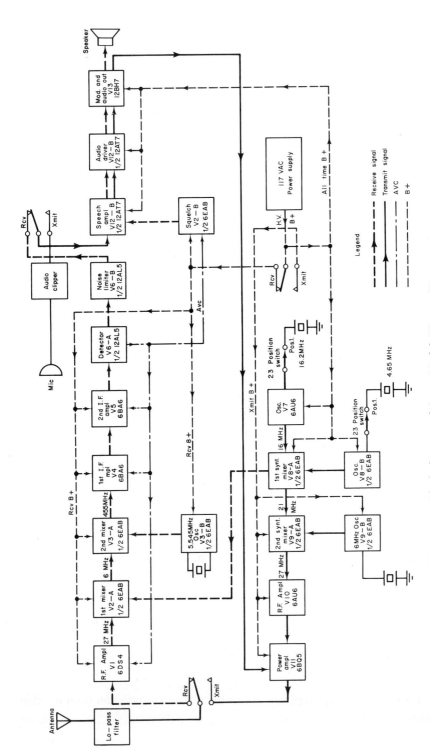

**Figure 1-4:** Block diagram of Sonar Model FS3023 CB transceiver

supply. A frequency synthesizer using 12 crystals determines the frequency of operation for both the transmitter and receiver sections. Output of the synthesizer is mixed with a 6 MHz oscillator output for transmitting or with a 5.545 MHz oscillator for receiving. A fine-tuning control is provided on the front panel so that the frequency of the 5.545 MHz oscillator may be shifted to accommodate the reception of transmitters operating within the tolerances provided by the FCC (0.005%).

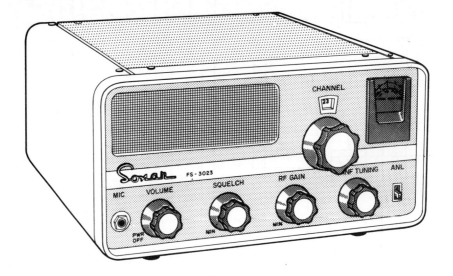

**Figure 1-5:** Sonar Model FS3023 CB transceiver

A channel indicator and meter are illuminated for better visibility. A large front-mounted speaker provides an ample amount of audio. A rear speaker jack is provided for an external speaker. The rear panel controls are adjusted during initial installation and do not need any attention, except during periodic maintenance. The meter reads both relative power output (of the transmitter section) and received signal strength (S-meter). A zero adjustment of the S-meter is provided on the rear panel.

The set is constructed of heavy-gauge aluminum for heavy-duty operation. The cabinet is properly ventilated. The perforated top and bottom are easily removed for servicing.

The transmitter is fully modulated (100%) by a Class B push-pull audio amplifier. The audio amplifers are designed to restrict the audio to 300–3000 Hz. This provides maximum intelligibility under poor operating conditions.

Sec. 1-2   Operation of a Typical CB Set                                    11

**1-2.2  Specifications**

The following specifications apply to the CB set shown in Figs. 1-4 and 1-5. However, the specifications are typical for many similar CB sets. The procedures for measurement of the specifications on typical CB sets are given throughout appropriate chapters of this handbook.

*Receiver specifications*

| | |
|---|---|
| Sensitivity $\frac{S+N}{N}$ | 0.4 $\mu$V for 10 dB |
|  | 1.0 $\mu$V for 20 dB |
| Selectivity | ±2.5 kHz at −6 dB |
|  | ±7.5 kHz at −60 dB |
| AGC | Output varies only 10 dB at inputs of 10 mV to 1 V |
| Squelch | adjustable to open at 0.5 to 300 $\mu$V |
| S-Meter | a reading to S9 denotes approximately 100 $\mu$V signal input |
| Noise limiter | gated series type (switched) |
| Audio output | at least 2.5 W at 10% distortion |
| Audio response | 300 to 3000 Hz at −6 dB |
| Hum and noise | at least −60 dB below full output |
| Image and spurious rejection | better than −60 dB |
| Adjacent channel rejection | better than −80 dB |

Note that all audio measurements are made with the noise limiter off. Sensitivity measurements are made with a 30% modulated signal source, with a modulating frequency of 400 Hz. Sensitivity measurements can vary from 4 to 6 dB due to production tolerances.

*Transmitter specifications*

| | |
|---|---|
| RF input power | 5 W (maximum permitted by FCC) |
| RF output power | 3.5 W |
| Modulation | not in excess of 100% |
| RF output impedance | 35 to 70Ω |
| Stability | better than 0.003% from −30 to +70°C and ±10% input voltage variation |

*General specifications*

| | |
|---|---|
| Primary power | 120 V ac at 96 W |
| Size (inches) | 11-3/4 wide by 5-3/4 high by 11-3/4 deep |
| Weight | 14 lb |

**1-2.3  Panel Controls and Indicators**

*Front panel controls and indicators.* (See Fig. 1-5.) These readily accessible controls and indicators are those most used during normal operation.

METER: Acts as an S-meter during receive (when the microphone push-to-talk button is not pressed). The meter acts as a *relative* RF power output meter during transmit (microphone button pushed).

CHANNEL SELECTOR: Switches any one of 23 channels. Rotation of the channel selector is continuous with no end stops from Channel 23 to 1 or from Channel 1 directly to 23.

FINE TUNING: Allows the operator to "peak up" any received signal that is not exactly on frequency.

MIC (microphone jack): Accepts any ceramic or carbon polarized microphone with a push-to-talk switch.

VOLUME Control: Determines the amount of audio output at the speaker. In the fully counter-clockwise position the CB set is turned off (PWR OFF).

SQUELCH: Restricts the background noise until a signal is received. Maximum squelch requires a large signal to open the threshold.

RF GAIN: Adjusts the overall sensitivity of the receiver. This is necessary for reception of very strong signals that may have a tendency to distort.

ANL (automatic noise limiter): **This switch controls the noise limiter circuit, which has been designed to reduce excessive electrical interference, ignition noise, etc. The automatic noise limiter function is disabled when the ANL switch is set to OFF.**

*Rear panel controls* (Not shown). These controls are not normally used during routine operation of the CB set.

EXTERNAL SPEAKER: This jack allows an external permanent magnet speaker of $4\Omega$ impedance to be plugged in. The front-mounted speaker is disabled when an external speaker is used.

ANTENNA: This is a UHF-type connector. Any antenna system with a nominal impedance of $50\Omega$ may be used.

ANTENNA LOADING: This control makes up for any variation in antenna impedance from 35 to $70\Omega$.

PLATE TUNING: This control resonates the RF power amplifier to provide maximum output from the transmitter to the antenna.

Note that both the *antenna loading* and *plate tuning* controls can affect operation of the transmitter (frequency, etc.). Thus, these controls must be adjusted by (or under the supervision of) a licensed operator, as discussed in Sec. 1-1.7.

Sec. 1-2  Operation of a Typical CB Set    13

S-METER ZERO: This control permits the front-panel S-meter to be zeroed (with the antenna disconnected, no signal applied, and the set operated in the receive condition). Since the control affects only the receiver circuits, anyone (licensed or unlicensed operators) can make this adjustment.

### 1-2.4  Basic CB Set Circuit Operation

The following descriptions apply to the circuits shown in block form in Fig. 1-4. However, the basic circuit arrangement is typical for millions of CB sets. The circuits of Fig. 1-4 can be considered as a composite for many sets. Detailed descriptions for similar circuits are given in Chapter 3.

*Microphone control circuits.* Once power is applied and the front-panel controls (VOLUME, SQUELCH, RF GAIN, FINE TUNING, and ANL) have been adjusted for best reception, operation of the set is controlled by the microphone push-to-talk button. When this button is pressed, power is applied to the coil of a relay, and four sets of relay contacts move to the energized or *transmit* position. The transmit position is shown as *xmit* on Fig. 1-4. Also note that only three sets of relay contacts are shown. The fourth set of contacts disconnects the speaker from the audio output $V_{13}$ when the contacts are in the transmit position.

One set of relay contacts connects the antenna (through a low-pass filter) to the output of the transmitter power amplifier $V_{11}$ when these contacts are in the *xmit* position. The same set of contacts connect the antenna to the input of the receiver RF amplifier $V_1$ in *rcv* position (when the microphone button is released, and power is removed from the relay coil).

Another set of contacts connects the microphone (through an audio clipper) to the input of the speed amplifier $V_{12A}$ in *xmit*. The same contacts connect the output of the receiver (at noise limiter $V_{6B}$) to the input of $V_{12A}$ in *rcv*. Thus, audio vacuum tubes $V_{12A}$, $V_{12B}$, and $V_{13}$ serve dual purposes. In *xmit*, the tubes function as speech amplifiers and modulators for the transmitter; they also serve to modulate power amplifier $V_{11}$ with voice signals impressed on the microphone. In *rcv*, the tubes function as audio amplifiers for the receiver and serve to amplify the detected audio signal to a level suitable for the speaker.

The remaining set of contacts apply B+ voltage to various transmitter circuits when the contacts are in *xmit* position and to the receiver circuits during *rcv*, thus energizing only the needed circuits.

Also note that B+ is applied to certain circuits of the set at all times, thus bypassing the relay contacts. These circuits are the audio section ($V_{12A}$, $V_{12B}$, $V_{13}$) and certain portions of the frequency synthesizer ($V_7$, $V_{8A}$, $V_{8B}$). This is necessary since these circuits are used for both receive and transmit. This emphasis on control circuit operation (however simple) may seem unnecessary to the experienced technician. However, a knowledge of how control circuits operate is vital when servicing any CB set.

*Receiver circuits.* The received signal from the antenna is fed through a low-pass filter and relay contacts to the input of $V_1$, which is a low-noise, neutralized triode amplifier. The output of $V_1$ is coupled to $V_{2A}$, where the signal is mixed with the output of the frequency synthesizer mixer $V_{8A}$. The frequency of the $V_1$ signal is approximately 27 MHz, whereas the $V_{8A}$ signal is approximately 21 MHz. This results in an output from $V_{2A}$ of about 6 MHz, which is coupled to $V_{3A}$. The 6 MHz signal is combined with a 5.545 MHz signal from crystal-controlled local oscillator $V_{3B}$ to produce an intermediate frequency (IF) signal of 455 kHz. Note that the crystal associated with $V_{3B}$ is variable. This is accomplished by connecting two diodes (that act as variable capacitors) across the crystal. The diode *capacitance* is varied by a voltage from the FINE TUNING control, thus shifting the crystal frequency slightly to either side of the fixed frequency. Note that the diode circuit details are not shown in Fig. 1-4. This presentation is typical for many, but not all, block diagrams shown in CB service literature.

The 455 kHz signal passes through a selective high-gain amplifier system consisting of $V_4$ and $V_5$. The amplified 455 kHz signal is demodulated (detected) by $V_{6A}$. The detected audio (300 to 3000 Hz) is applied through gated noise limiter $V_{6B}$, the VOLUME CONTROL, and the relay contacts (*in rcv*) to the input of $V_{12A}$. Note again that the VOLUME CONTROL is not shown in Fig. 1-4, nor is the RF GAIN control, which sets voltage on the cathodes of receiver tubes $V_1$ and $V_4$ (thus setting the sensitivity of both the RF and IF sections).

*Transmitter circuits.* The frequency of the transmitter section is controlled by the frequency synthesizer circuits. Practically all present-day CB sets use some form of synthesizer. Such circuits make it possible to control 23 or 40 channels (of both transmission and reception) with a minimum of crystals (thus reducing size, weight, and cost). Without a frequency synthesizer, it would require 23 crystals for 23 channels of operation. The circuit of Fig. 1-4 uses 12 crystals to control the frequency of transmission and reception on all 23 channels.

The crystal that controls receiver oscillator $V_{3B}$ is fixed at 5.545 MHz. The crystal that controls transmitter oscillator $V_{9B}$ is at 6 MHz. Four crystals are used to control oscillator $V_{8B}$. The frequencies and corresponding channels are given in Fig. 1-6. Six crystals control oscillator $V_7$. The frequencies and channels are also given in Fig. 1-6.

Any channel can be covered by proper combinations of the crystals.

Sec. 1-2  Operation of a Typical CB Set

**4 MHz OSCILLATOR**

| Crystals (MHz) | Channels |
|---|---|
| 4.765 | 1, 5, 9, 13, 17, 21 |
| 4.775 | 2, 6, 10, 14, 18, 22 |
| 4.785 | 3, 7, 11, 15, 19 |
| 4.805 | 4, 8, 12, 16, 20, 23 |

**16 MHz OSCILLATOR**

| Crystals (MHz) | Channels |
|---|---|
| 16.200 | 1 thru 4 |
| 16.250 | 5 thru 8 |
| 16.300 | 9 thru 12 |
| 16.350 | 13 thru 16 |
| 16.400 | 17 thru 20 |
| 16.450 | 21 thru 23 |

**Figure 1-6:** Crystal frequencies used in the frequency synthesizer

For example, assume that it is desired to transmit on channel 3 (26.985 MHz). When the CHANNEL SELECTOR is set to 3, the 4.785 MHz crystal is selected for $V_{8B}$, and the 16.200 MHz crystal is selected for $V_7$. This produces an output of 20.985 MHz from $V_{8A}$, which is mixed with the 6 MHz signal from $V_{9B}$ in mixer $V_{9A}$. This produces the desired output of 26.985. If channel 7 is selected, $V_{8B}$ remains at 4.785, but $V_7$ changes to 16.250, producing an output of 21.035 MHz. This is mixed with the 6 MHz signal to produce the desired 27.035 MHz.

Now assume that you want to receive on channel 7. The incoming signal is still 27.035 MHz, which is mixed with the 21.035 MHz signal from $V_{8A}$ to produce a 6 MHz signal in $V_{2A}$. The output of $V_{2A}$ is combined with the 5.545 MHz signal from $V_{3B}$ to produce a 455 kHz signal.

During transmission, the output of synthesizer mixer $V_{9A}$ is applied to $V_{10}$. RF amplifier $V_{10}$ raises the level of the 27 MHz signal to drive RF power amplifier $V_{11}$ to full output (3.5 W). $V_{11}$ is modulated by the output of $V_{13}$. Amplifier $V_{11}$ is Class C, and thus produces many harmonics. These harmonics can interfere with other radio communications services (TV, aviation, mobile, etc.). The low-pass filter shown in Fig. 1-4 passes signals below about 30 MHz (thus covering all of the CB channels) but rejects signals above 30 MHz. Most present-day CB sets have some form of low-pass filter. Likewise, there are external low-pass filter units available for older CB sets not so equipped.

*Audio output and modulator circuit.* As explained previously, tubes $V_{12}$ and $V_{13}$ serve a dual purpose. In transmit, $V_{12A}$ is a microphone speech amplifier, $V_{12B}$ is an audio driver, and $V_{13}$ is a modulator for $V_{11}$. In receive, $V_{12A}$ is the receiver speech amplifier (or first audio amplifier), $V_{12B}$ is an audio driver (or second audio amplifier), and $V_{13}$ is the final or output audio amplifier for the speaker. In receive, the threshold of $V_{12A}$ (and thus the entire audio section) is set by squelch circuit $V_{2B}$. In turn, $V_{2B}$ is controlled by the front-panel SQUELCH control. This makes it possible to set the threshold of the audio section to prevent low-level noise from passing to the speaker, but permits higher-level voice signals to pass.

*Automatic volume control (AVC) circuits.* Note that $V_1$, $V_4$, $V_5$, and $V_{2B}$ receive AVC signals from the detector $V_{6A}$. The amplitude of the AVC signals depends on the amplitude of the incoming signals. If the incoming signals are strong, the AVC signals bias the corresponding circuits so as to reduce sensitivity. Weak incoming signals produce the opposite effect. Thus, the audio output remains constant (within limits) in the presence of received signals that vary in strength. When no signal is received and no AVC voltage is present, $V_{2B}$ can be adjusted until $V_{12A}$ will not pass audio of any given level. When a signal above the selected level is received and AVC voltage is applied to $V_{2B}$, $V_{12A}$ then passes audio to $V_{12B}$.

*Audio clipper circuits.* These circuits, mounted on a printed circuit board in this case, prevent modulation in excess of 100% on either positive or negative peaks. The clipper circuits are adjustable to set a desired level of modulation for a given voice input.

## 1-3 THE BASIC TROUBLESHOOTING FUNCTIONS

Troubleshooting can be considered a step-by-step logical approach to locating and correcting any fault in the operation of a piece of equipment. In the case of CB troubleshooting, seven basic functions are required.

First, you the technician must study the CB set, using service literature, schematic diagrams, and so forth, to find out how each circuit works when it is operating normally. In this way, you will know in detail how a given set should work. If you do not take the time to learn what is normal operation, you will never be able to distinguish what is abnormal.

Second, you must know the function of all controls, indicators, and adjustments, and how to manipulate them. It is difficult to check out a set without knowing how to use the controls, even though CB set controls are fairly simple and relatively standard. Also, as the set ages, readjustment and realignment of critical circuits may be required.

## Sec. 1-3 The Basic Troubleshooting Functions

Third, you must know how to interpret service literature and how to use test equipment. Along with good test equipment that you know how to use, well-written service literature is your best friend.

Fourth, you must be able to apply a systematic, logical procedure in order to locate the trouble. Of course, a procedure that is logical for one type of set is quite illogical for another. For that reason, we discuss logical troubleshooting procedures for various sets, as well as basic procedures that apply to all sets.

Fifth, you must be able to analyze logically the information provided by an improperly operating set. The information to be analyzed may be the set's performance (transmission and reception on all channels), indications taken from test equipment (voltage and resistance measurements), or indications taken from the set's front panel indicators (received signal strength and/or transmitted RF power). No matter what form the information takes, it is your analysis of the information that makes for logical, efficient troubleshooting.

Sixth, you must be able to perform complete checkout procedures on the repaired set. Such a checkout may require only simple operation (switching through all channels, checking transmission and reception, squelch, ANL, and volume control operation). At the other extreme, the checkout may involve complete realignment of the set. Either way, a checkout is always recommended after troubleshooting. One reason for a checkout is that there may be more than one problem. For example, an aging part may cause high current to flow through a resistor, resulting in the burnout of the resistor. Logical troubleshooting may lead you quickly to the burned-out resistor. Replacement of the resistor will restore operation. However, only a thorough checkout will reveal the original high-current condition that caused the burnout.

Another reason for after-service checkout is that the repair may have produced a condition that requires readjustment. A classic example of this occurs when replacement of a part changes circuit characteristics. For example, a new transistor in an RF or IF stage may require complete realignment of the stage.

Seventh, you must be able to use the proper tools to repair the trouble. As a minimum for CB repair, you must be able to use soldering tools, wire cutters, long-nose pliers, screwdrivers, and socket wrenches. If you are still at the stage where any of these tools seem unfamiliar, you are not ready for CB service, even simpled service.

In summary, before starting any CB service work, ask yourself these questions: Have I studied all available service literature to find out how the particular CB set works? Can I operate the set properly? Do I really understand the service literature, and can I use all required test equipment properly? Using the service literature and/or previous experience on similar equipment, can I plan a logical troubleshooting procedure?

Can I analyze the results of operating checks, as well as procedures involving test equipment, logically? Using the service literature and/or experience, can I perform complete checkout procedures on the set, including realignment, adjustment, and so forth, if necessary? Once I have found the trouble, can I use common hand tools to make the repairs? If the answer to any of these questions is no, you are simply not ready for CB service. Start studying instead!

## 1-4. THE UNIVERSAL TROUBLESHOOTING PROCEDURE

The troubleshooting functions of CB service discussed thus far may be divided into four major steps:

1. *Determine* the trouble symptoms.
2. *Localize* the trouble to a functional area.
3. *Isolate* the trouble to a circuit.
4. *Locate* the specific trouble, probably to a specific part.

The remaining sections of this chapter are devoted to these four steps. Before going into the details of the steps, let us examine what is accomplished by each.

### 1-4.1 Determining Trouble Symptoms

Determining symptoms means that you must know what the set is supposed to do when it is operating normally, and, in addition, you must be able to recognize when that normal job is not being done. Most CB sets have operating controls, indicating instruments, or other built-in aids for evaluating their performance. As a minimum, there is a loudspeaker (to indicate voice reception). Generally, there is also a meter (such as shown in Fig. 1-5) to indicate received signal strength (S-meter) and transmitted power (relative RF power output). Such meters are commonly known as S/RF or RF/S meters. Also, most CB sets will have a channel selector, on-off-volume control, and probably a squelch control.

You must analyze the normal and abnormal symptoms produced by the set's built-in indicators in order to formulate the following questions: What is the set supposed to do (does it cover 23 or 40 channels, etc.)? How well is this job being done (is transmission and/or reception normal on any, all, or only certain channels)? Where in the set could there be trouble that will produce these symptoms?

The "determining-symptoms" step does not mean that you should charge into the set with screwdriver and soldering tool, nor does it mean

Sec. 1-4 The Universal Troubleshooting Procedure 19

that test equipment should be used extensively. Instead, it means that you make a visual check, noting both normal and abnormal performance indications. It also means that you operate the controls to gain further information.

At the end of the "determining-symptoms" step, you know definitely that something is wrong and have a fair idea of what is wrong, but you probably do not know just what area of the set is faulty. This is established in the next step of troubleshooting.

### 1-4.2 Localizing Trouble to a Functioning Area

Most CB sets may be subdivided into areas that have a definite purpose or function. The term *function* is used in CB troubleshooting to mean an operation performed in a specific area of the set. For example, the set shown in Fig. 1-4 may be divided into receiver, transmitter, power supply, crystal control (frequency synthesizer), transmit-receive/control (switching relay), audio, and low-pass filter.

To localize the trouble systematically and logically, you must have a knowledge of the functional areas and must correlate all the symptoms previously determined. Thus, you might first determine, by an educated guess, the functional area most likely to cause the indicated symptoms. First off, you may consider several technically accurate possibilities as the probable trouble area.

As an oversimplified example, if modulation is poor during transmission, and voice is weak during reception, the audio section ($V_{12}$ and $V_{13}$) is a likely suspect, since it is common to both transmission and reception. On the other hand, if the transmission is good but reception is poor, the trouble is probably in the receiver section ($V_1$ through $V_6$), since these circuits apply only to reception.

*Use of diagrams.* CB troubleshooting involves (or should involve) the extensive use of diagrams. Such diagrams may include a *functional block diagram* and almost always include *schematic diagrams*. (*Practical wiring diagrams*, such as are found in military-type service literature, are almost never available for CB service. Most CB set parts are mounted on a printed circuit (PC) board or boards. At best, the service literature will show the PC board layout, and possibly the wiring from the boards to the external controls, connectors, and indicators.)

The block diagram (such as shown in Fig. 1-4) illustrates the functional relationship of all *circuits* in the set, and is thus the most logical source of information for trouble localization. Unfortunately, not all CB service literature is provided with a block diagram. It may be necessary to use the schematic diagram.

The schematic diagram (such as shown in Chapter 3) shows the functional relationship of all *parts* in the set. Such parts include all transistors, vacuum tubes, capacitors, transformers, diodes, etc. Generally, the schematic presents too much information (not directly related to the specific symptoms noted) to be of maximum value during the localizing step. The decisions made regarding the probable trouble area may become lost among all the details. However, the schematic is very useful in later stages of the total troubleshooting effort, or when a block diagram is not available.

In comparing the block diagram and the schematic during the localizing step, note that each transistor or vacuum tube shown on the schematic is usually represented as a block on the block diagram. This relationship is typical on most CB service literature.

The physical relationship of parts is often given on *component location diagrams* (also called *parts placement* or *parts identification diagrams*). These location or placement diagrams rarely show or identify all parts, as do true military-style practical wiring diagrams. Instead, the parts location diagrams concentrate on identification of major parts such as transistors and vacuum tubes, transformers, diodes, and adjustment controls. For this reason, location diagrams are least useful in localizing trouble. Instead, the location diagrams are most useful in locating specific parts during other phases of troubleshooting.

To sum up, it is logical to use a block diagram instead of a schematic or location diagram when you want to make a good guess as to the probable trouble areas. The use of a block diagram also permits you to use a troubleshooting technique known as *bracketing* (discussed in Sec. 1-7). If the block diagram includes major test points, as it may in some well-prepared service literature, the block will also permit you to use test equipment as aids in narrowing down the probable cause of trouble. However, test equipment is used more extensively during the isolation step of troubleshooting.

### 1-4.3 Isolating Trouble to a Circuit

After the trouble is localized to a single functional area, the next step is to isolate the trouble to a circuit in the faulty area. To do this, concentrate on those circuits in the area that could cause the trouble and ignore the remaining circuits.

The isolating step involves the use of a test equipment such as meters, oscilloscopes, and signal generators for *signal tracing* and *signal substitution* in the suspected faulty area. By making educated estimates and properly using the applicable diagrams, bracketing techniques, signal tracing, and signal substitution, you can systematically and logically isolate the trouble to a single defective circuit.

Sec. 1-4 The Universal Troubleshooting Procedure 21

Repair techniques or tools to make necessary repairs to the set are not used until after the specific trouble is located and verified. That is, you *still* do not charge into the set with solder tools and screwdriver at this point. Instead, you are now trying to isolate the trouble to a specific defective circuit so that, once the trouble is located, it can be repaired.

### 1-4.4 Locating the Specific Trouble

Not only does this troubleshooting step involve locating the specific trouble, it also includes a final analysis, or review, of the complete procedure, as well as the use of repair techniques to remedy the trouble once it has been located. This final analysis will permit you to determine whether some other malfunction caused the part to be faulty or whether the part located is the actual cause of the trouble.

Inspection by using the senses—sight, smell, hearing, and touch—is very useful in trying to locate the trouble. This inspection is usually performed first, in order to gather information that may more quickly lead to the defective part. (The inspection is often referred to as a "visual inspection" in service literature, although it involves all of the senses.) Among other things to look for during "visual inspection" are burned, charred, or overheated parts, arcing in the circuits, and burned-out parts.

In sets where access to the circuitry is relatively easy, a rapid visual inspection should be performed first. Then the active device—vacuum tube or transistor—can be checked. A visual inspection is always recommended as the first step in all solid-state sets and in most vacuum-tube sets. A possible exception is sets in which access to circuit parts is very difficult, but where vacuum tubes can be easily removed and tested (or substituted). Vacuum tube testing is discussed further in Sec. 1-9.

The next step in locating the specific trouble is the use of an oscilloscope to *observe waveforms* and a meter to *measure voltages*. The oscilloscope is not used as extensively in CB service as in other fields of electronic troubleshooting (such as television service), because most of the waveforms in CB are RF, IF, and audio signals, rather than pulses (as in TV circuits). However, an oscilloscope should be used to measure the modulation envelope of a CB set; it may also be used as a substitute meter for voltage measurements. Of course, a conventional meter is best when making *resistance* and *continuity* checks to pinpoint a defective part. After the trouble is located, you should make a final analysis of the complete troubleshooting procedure to verify the trouble. Then you can repair the trouble and check out the set for proper operation.

Note that in most CB service literature the voltages (and possibly the resistances) are often given on the schematic diagram, but this information may also appear in chart form (following the military style). Except possibly for a diagram of the modulation envelope, waveforms are rarely found in present-day CB service literature. No matter what information is given, and what form it may take, you must be able to use test equipment to make the measurements. For that reason, the function and use of test equipment during troubleshooting is discussed frequently throughout this handbook.

### 1-4.5 Developing a Systematic, Logical Troubleshooting Procedure

The development of a systematic and logical troubleshooting procedure requires:

A logical approach to the problem

Knowledge of the set

Interpretation of test information

The use of information gained in each step

Some CB technicians feel that a knowledge of the set involves remembering past failures as well as such things as the location of all test points, all adjustment procedures, etc. This approach may be good in troubleshooting only one type of set, but it has little value in developing a basic troubleshooting procedure.

It is true that recalling past set failures may be helpful, but you should not expect that the same trouble will be the cause of a given symptom in every case. In any CB set, many trouble areas may show approximately the same symptom indications.

Also, you should never rely only on your memory of adjustment procedures, test point locations, etc., in dealing with any troubleshooting problem. This is one of the functions of service literature containing diagrams and information on CB sets. Chapter 6 discusses the specific use of, and the types of information to be found in, service literature. The important point for you to learn is to be a systematic, logical troubleshooter, not a memory expert.

### 1-4.6 Relationship among Troubleshooting Steps

Thus far, we have established the overall troubleshooting approach. Now let us make sure you understand how each troubleshooting step fits together with the others by analyzing their relationships.

The first step, *determining the symptoms*, requires the use of the senses, the observation of set performance, previous knowledge of set operation, the manipulation of operating controls, and possibly the

Sec. 1-4   The Universal Troubleshooting Procedure                          23

recording of notes. Determining the symptoms presupposes the ability to recognize inappropriate indications, to operate the controls properly, and to note the effect that the controls have on trouble symptoms.

The second step, *localizing the trouble to a functional area*, depends on the information gained in the first step, plus the use of a functional block diagram (or possibly the schematic), and reasoning. During the second step, ask yourself the question: What functional area could cause the indicated symptoms? Then bracketing, or narrowing down the probable defect to a single functioning area, is used along with the test equipment to actually pinpoint the faulty function. The observations in this step depend on noting the indications of the testing devices used to localize the trouble.

The third step, *isolating the trouble to a circuit*, uses all of the information gathered up to this time. The main difference between this step and the second step is that now schematic diagrams are used instead of the block diagrams, and test equipment is used extensively.

In the fourth step, *locating the trouble*, all the findings are reviewed and verified to ensure that the suspected part is the cause of the failure. This final step also includes the necessary repair (replacement of defective parts, etc.) as well as a final checkout.

*Example of a relationship among troubleshooting steps.* To make sure that you understand the relationships of the troubleshooting steps, let us consider an example. Assume that you are troubleshooting a set, you are well into the fourth step ("locate"), and you find nothing wrong with the circuit (for example the receiver circuit). That is, all waveforms, voltage measurements, and resistance measurements are normal. What is your next step?

You might assume that nothing is wrong—that the problem is "customer trouble." This is poor judgment. First, there must be something wrong in the set since some abnormal symptoms were recognized before you got to the locate step. Never assume anything when troubleshooting; either the set is working properly, or it is not working properly; either observations and measurements are made, or they are not made. You must draw the right conclusions from the observations, measurements, and other factual evidence, or you must repeat the troubleshooting procedure.

*Repeating the troubleshooting procedure.* Some technicians new to service work assume that repeating the troubleshooting procedure means starting all over from the first step, and, in fact, some service literature recommends this action, since it is possible for anyone, even an experienced technician, to make mistakes. When performed logically and systematically, the troubleshooting procedure will keep mistakes to a minimum. However, voltage and resistance measurements may be interpreted erroneously, waveform observations or bracketing may be performed incorrectly, or many other mistakes may occur through simple oversight.

In spite of such recommendations by other service literature writers, this author contends that "repeat the troubleshooting procedure" means *retrace* your steps, one at a time, until you find the place where you went wrong. Perhaps a previous voltage or resistance measurement was interpreted erroneously in the locate step, or perhaps a waveform observation or bracketing step was incorrectly performed in the isolate step. The cause must be logically and systematically determined by taking a *return path* to the point at which you went astray.

## 1-5 APPLYING THE TROUBLESHOOTING APPROACH TO CB SERVICE

Now that we have reviewed the basic operating principles of CB and have established a basic troubleshooting approach, let us discuss how this approach can be applied to the specifics of CB service. The remainder of this chapter is devoted to generalized examples of how the approach may be used to troubleshoot CB sets. Detailed descriptions of CB troubleshooting examples are given in Chapter 4.

## 1-6 TROUBLE SYMPTOMS

It is impractical to list all the trouble symptoms that may occur in all CB sets. However, the list in Fig. 1-7 covers those problems most commonly encountered in a typical CB set. The troubles are grouped into functional areas (or circuits) of the set. These areas or circuit groups correspond to those of the block diagram in Fig. 1-4 and the discussions of Sec. 1-2.

Some of the symptoms listed point to only one area of the set as a *probable cause* of trouble. For example, if there is no reception on any channel, but transmission is normal on all channels, the trouble is *most likely* in the receiver circuits. On the other hand, if there is some problem in both transmission and reception, the trouble *could be* in the low-pass filter, the audio and modulator circuits, the relay switching circuits, or possibly in the power supply.

In the discussions of the localize, isolate, and locate steps, we gave examples of how the symptoms could be used as the first step in pinpointing trouble to an area, to a circuit within the area, and finally to a part within the circuit. Before discussing these steps further, let us note ways of dealing with symptoms of CB set troubleshooting.

Sec. 1-6 Trouble Symptoms

### 1-6.1 Determining Trouble Symptoms

It is obvious that trouble exists when electronic equipment will not operate, for example, when a CB set is connected to power, turned on, the controls are properly set, but there is no transmission, reception, and the panel lamp is off. A different problem exists when the equipment is still operating but is not doing its job properly. Using the same CB set, assume that transmission and reception are present but poor.

Another difficulty in determining trouble symptoms is improper use of the equipment by the operator. In complex electronic equipment, operators are usually trained and checked out on the equipment. The opposite is true of CB sets used by the general public. However, no matter what equipment is involved, it is always possible for an operator (or customer) to report a "trouble" that is actually a result of improper operation. For these reasons, you must first determine the *signs* of failure, regardless of how bad it may appear to be, and without regard to the cause (set failure or operator trouble). This means that you must know how the set operates normally, and how to operate the controls.

### 1-6.2 Recognizing Trouble Symptoms

Symptom recognition is the art of identifying *normal* and *abnormal* signs of operation in electronic equipment. A trouble symptom is an undesired *change* in equipment performance or a deviation from the standard. For example, the RF output indicator (panel meter) should show some RF output when the push-to-talk button is pressed. If there is no RF indication, you should recognize this as a trouble symptom, because it does not correspond to the normal, expected performance.

Now assume that the same CB set has poor reception, perhaps due to bad signal conditions in the area, or a defective antenna. If the receiver circuits of the set do not have sufficient gain to produce good reception under these conditions, you could mistake this for a trouble symptom, unless you were really familiar with the set. Poor reception (for this particular set operating under these conditions) is not abnormal operation, nor is it an undesired change. Thus, it is not a true trouble symptom and should be so recognized.

### 1-6.3 Equipment Failure versus Degraded Performance

Equipment failure means that either the entire equipment or some functional part of the equipment is not operating properly. For example, the total absence of any received signal when all controls are properly set is a form of equipment failure, even though there may be sound (background noise) from the loudspeaker. Degraded performance occurs whenever the equipment is working but is not presenting normal performance. For example, the presence of hum in the loudspeaker is degraded performance, since the set has not yet failed but the performance is abnormal.

### 1-6.4 Evaluation of Symptoms

Symptom evaluation is the process of obtaining more detailed descriptions of the trouble symptoms. The recognition of the original trouble may not in itself provide enough information to decide on the probable cause or causes of the trouble, because many faults produce similar trouble symptoms.

To evaluate a trouble symptom, it is generally necessary to operate the controls associated with the symptom and apply your knowledge of electronic circuits, supplemented with information gained from the service literature. Of course, the mere adjustment of operating controls is not the complete story of symptom evaluation. However, the discovery of an incorrect setting can be considered a part of the overall symptom evaluation process.

### 1-6.5 Example of Evaluating Symptoms

When there is no sound of any kind coming from the loudspeaker of a CB set, there obviously is trouble. The trouble could be caused by a shorted transistor, burned-out diode, defective capacitor, or any one of the several hundred components in the RF, IF, and audio circuits (assuming that power is applied, the set is switched on, and the squelch is properly set). However, the same symptom may be produced when the RF gain control is turned down (minimum sensitivity). Think of all the time you may save by checking the operating controls first, before you charge into the set with tools and test equipment!

To do a truly first-rate job of determining trouble symptoms, you must have a complete and thorough knowledge of the normal operating characteristics of the set. Your knowledge helps you decide if the set is doing the job for which it was designed. In most service literature this is more properly classified as "knowing your equipment."

In addition to knowing the set, you must be able to operate all the controls properly in order to determine the symptom—to decide on *normal* or *abnormal* performance. If the trouble is cleared up by manipulating the controls, your analysis may or may not stop at this point. Through your knowledge of the set, you should be able to understand why a specific control adjustment removed the apparent trouble.

## 1-7 LOCALIZING TROUBLE

*Localizing trouble* means that you must determine which of the major functional areas in a set are actually at fault. This is done by systematically checking each area selected until the faulty one is found. If none of the functional areas on your list shows improper performance, you must take a return path and recheck the symptom information (and observe more information, if possible). Several circuits could be causing the trouble, and the localize step will narrow the list to those in one functional area, as indicated by a particular block of the block diagram.

The problem of trouble localization is simplified when a block diagram and a list of trouble symptoms (such as those shown in Figs. 1-4 and 1-7) are available for the CB set being serviced. Keep in mind that

POWER SUPPLY

    Set dead, no light in channel indicator or meter.
    No transmission, no reception on any channel.

RECEIVER ($V_1$ THROUGH $V_6$)

    No reception on any channel, transmission normal.
    Reception poor, transmission good.

TRANSMITTER ($V_9$, $V_{10}$, $V_{11}$)

    No transmission on any channel, reception normal.
    Transmission poor, reception good.

AUDIO AND MODULATION CIRCUITS ($V_{12}$, $V_{13}$)

    No modulation, but carrier present (on RF output meter).
    No sound in speaker with volume control full on.

AUDIO CLIPPER

    Poor modulation, sound good on reception.

FREQUENCY SYNTHESIZER ($V_7$, $V_8$)

    Transmission and reception abnormal on certain channels.
    Transmission off-frequency.

CONTROL CIRCUITS (SWITCHING RELAY)

    No transmission with push-to-talk switch pressed.

LO-PASS FILTER

    Poor reception and transmission, carrier normal on output meter.

**Figure 1-7:** Typical CB set trouble symptoms

these illustrations apply to a "typical" or composite set. However, the general arrangement shown in Figs. 1-4 and 1-7 can be applied to many sets. Thus, the illustrations serve as a universal starting point for trouble localization.

### 1-7.1 Bracketing Technique

The basic bracketing technique makes use of a block diagram or schematic to localize the trouble to a functional area. Bracketing (sometimes known as the *good input/bad output* technique) provides a means of narrowing the trouble down to a circuit group and then to a faulty circuit. Symptom analysis and/or signal-tracing tests are used in conjunction with, or are a part of, bracketing.

Bracketing starts by placing brackets (at the good input and the bad output) on the block diagram or schematic. Bracketing can be done mentally, or it can be physically marked with a pencil, whichever is most effective for you. No matter what system is used, with the brackets properly positioned, you know that the trouble exists somewhere between the two brackets.

The technique involves moving the brackets, one at a time (either the good input or the bad output), and then making tests to find if the trouble is within the newly bracketed area. This process continues until the brackets localize a circuit group.

The most important factor in bracketing is to find where the brackets should be moved in the elimination process. This is determined from your deductions based on your knowledge of the set and on the symptoms. All moves of the brackets should be aimed at localizing the trouble with a minimum of tests.

### 1-7.2 Examples of Bracketing

Bracketing may be used with or without actual measurement of voltages or signals. That is, sometimes localization can be made on the basis of symptom evaluation alone. In practical CB service, both symptom evaluation and tests are usually required, often simultaneously. The following examples show how the technique is used in both cases.

Assume that you are servicing a CB set and that you find a "no-reception and no-transmission" symptom. That is, the power is applied, the set is turned on, the pilot lamps are on, but there is no RF indication on the panel meter during transmission and no signal strength indication during reception. The power supply is a logical suspect as the faulty circuit group.

You place a good-input bracket at the 115 V input, and a bad-output bracket at the dc output, as shown in Fig. 1-8. To confirm the symptom, you measure both the dc output voltage (or voltages) and the ac input

Sec. 1-7 Localizing Trouble 29

voltages. If the input is normal but one or more of the output voltages is absent or abnormal, you have localized the trouble to the power supply circuits. The next step is to isolate the trouble to a specific circuit in the power supply, as discussed in Sec. 1-8.

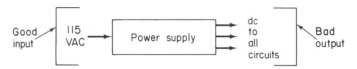

**Figure 1-8:** Example of bracketing for "set dead" symptom

From a practical troubleshooting standpoint, it is possible that the power supply output voltages are normal, but you still have a "no-reception and no-transmission" symptom. For example, the lines carrying the dc voltages to other circuit groups could be open, or the switching relay circuits could be malfunctioning to interrupt the dc voltages. This can be checked by measuring the voltages at the circuit end of the lines as well as at the power supply end.

Also, it is possible that a failure in another circuit could cause the power supply output voltage to be abnormal. For example, if there is a short in one of the circuits on the dc supply line, the dc output voltage will be low. Of course, this will show up as an abnormal measurement and will be tracked down during the "isolate" step of troubleshooting.

As another example of bracketing, now assume that the "no-reception and no-transmission" condition still exists, but the symptoms are somewhat different. Now, there is an RF indication on the panel meter during transmission, and a signal strength indication during reception, but there is no sound in the loudspeaker. You could start by placing a good-input bracket at the input to the audio and modulation circuits and a bad-output bracket at the output of these circuits (loudspeaker and/or modulation transformer), as shown in Fig. 1-9. However,

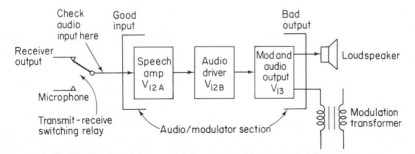

**Figure 1-9:** Example of bracketing for "no-reception, no-transmission, good S/RF meter indications" symptom

from a practical standpoint, your first move should be adjustment of the volume and squelch controls.

If the trouble is not cleared by adjustment of the controls, confirm the good-input bracket by monitoring the signal at the audio and modulator circuit input (during reception there should be audio at this point). Make this check at the input to the audio, as shown in Fig. 1-9. Possibly the line between the receiver and audio circuits is open, or perhaps the switching relay contacts are defective. It is also possible that the line is partially shorted. (A completely shorted line would probably cause failure of the receiver circuits and could result in a lack of signal indication on the panel meter.)

If there are audio signals at the input of the audio and modulator circuits, but there is no sound on the loudspeaker (even with adjustment of the volume and squelch controls), you have localized the trouble to the audio and modulator circuits. The next step is to isolate the trouble to a specific circuit in the audio and modulator group, as discussed in Sec. 1-8.

### 1-7.3 Localization with Replaceable Modules

The localization procedure can be modified when the circuits of a CB set are located on replaceable modules. The trend in present-day CB sets is toward the use of replaceable modules, such as PC (printed-circuit) boards that are either plug-in or require only a few soldered connections. In such sets it is possible to replace each module or board in turn until the trouble is cleared. For example, if replacement of the audio module restores normal operation, the defect is in the audio module. This conclusion may be confirmed by reinserting the suspected defective module. Although this confirmation process is not a part of theoretical troubleshooting, it is a good practical check, particularly in the case of a plug-in module. Often a trouble symptom of this sort may be the result of a poor contact between the plug-in module and the chassis connector or receptacle.

Some service literature recommends that tests be made before all modules are arbitrarily replaced, usually because the modules are not necessarily arranged according to function area. Thus, there is no direct relationship between the trouble symptom and the modules. In such cases, always follow the service literature recommendations. Of course, if modules are not readily available in the field, you must make tests to localize the trouble to a module (so that you can order the right module, for example). Also, operation controls and connectors are not usually found on replaceable modules, so they must be tested separately.

Sec. 1-7 Localizing Trouble

### 1-7.4 Which Circuit Groups to Test First

When you have localized trouble to more than one circuit group you must decide which group to test first. Several factors should be considered in making this decision.

As a rule, if you can *run a test that eliminates several circuits*, or circuit groups, use that test first, before making a test that eliminates only one circuit. This requires an examination of the diagrams (block and/or schematic) and a knowledge of how the set operates. The decision also requires that you apply logic.

*Test point accessibility* is the next factor to consider. A test point can be a special jack located at an accessible spot (say at the top of the chassis). The jack (or possibly a terminal) is electrically connected (directly or by a switch) to some important operating voltage or signal path. At the other extreme, a test point can be *any point* where wires join or where parts are connected together.

Another factor (although definitely not the most important) is your past experience and a history of *repeated set failures*. Past experience with identical or similar sets and related trouble symptoms, as well as the probability of failure based on on records of repeated failures, should have some bearing on the choice of a first test point. However, all circuit groups related to the trouble symptom should be tested, no matter how much experience you may have had with the set. Of course, the experience factor may help you decide which group to test first.

Anyone who has had any practical experience in troubleshooting knows that all the steps of a localization sequence rarely proceed in textbook fashion. Just as true is the fact that many troubles listed in the CB set service literature may never occur in the set you are servicing. These troubles are included in the literature as a guide, and are not meant to be hard and fast rules. In some cases of localizing the trouble it may be necessary to modify your troubleshooting procedure. The physical arrangement of the set may pose special troubleshooting problems. Also, special knowledge gained from experience with similar sets may simplify the task of localizing the trouble.

### 1-7.5 Universal Trouble Localization

In the paragraphs below, we shall describe a universal trouble localization process for a typical CB set. The procedures are based on the assumption that the set circuit arrangement is as shown in Fig. 1-4; thus it is possible to group the troubles as shown in Fig. 1-7. The details of these localization procedures are elaborated in the remaining sections of this chapter and in Chapter 4.

**32**  Ch. 1  INTRODUCTION TO CB SERVICE

*If the set is completely "dead"* (no panel lights, no transmission, no reception), check the input and output of the power supply. Also check the power supply fuse (if any). If one or more power supply output voltages are absent or abnormal, you have localized the problem to the power supply circuits. If the power supply output voltages are normal, check for proper distribution of voltages at the remaining circuits (receiver, transmitter, etc.). If the receiver voltages are normal, but not the transmitter (or vice versa), check the relay switching circuits.

*If there is no reception, or reception is poor,* but transmission is normal (with all controls properly set), check for a signal at the receiver output with the receiver tuned to an active channel. The volume control is generally a convenient test point, as shown in Fig. 1-10. The audio

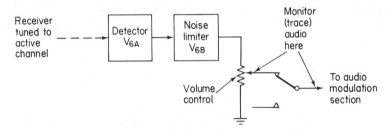

**Figure 1-10:** Monitoring audio output of receiver to trace a "no-reception or reception poor" symptom

signal can be monitored (traced) at this point (to check operation of the receiver circuits $V_1$ through $V_6$). As an alternate, an audio signal can be injected at the volume control (to check operation of the audio and modulation section $V_{12}$ through $V_{13}$), as shown in Fig. 1-11. These same

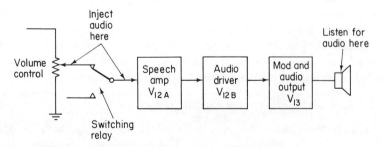

**Figure 1-11:** Injecting audio at receiver output to clear audio/modulation section

tests can be made at the relay contacts that switch the audio and modulation circuit input between the receiver output and audio clipper. Either way, if an audio signal is present at the volume control or relay contacts,

Sec. 1-7  Localizing Trouble 33

the receiver circuits are cleared. Using the alternate test, if the audio signal passes to the speaker, the audio and modulator circuits are cleared.

Instead of signal tracing through the receiver circuits, the receiver can also be checked by injecting a modulated RF signal at the antenna jack (the RF signal must be modulated by an audio tone) and listening for the modulated tone in the speaker, as shown in Fig. 1-12. Since this symptom may be caused by a defect in the switching control circuits, inject the modulated RF signal on both sides of the relay contacts as shown.

To eliminate the AVC circuits as trouble suspects, apply a fixed dc voltage to the AVC line and check operation of the receiver. (This is known as *clamping* the AVC line). If operation is normal with the AVC line clamped but not when the clamp is removed, you have localized trouble to the AVC circuit.

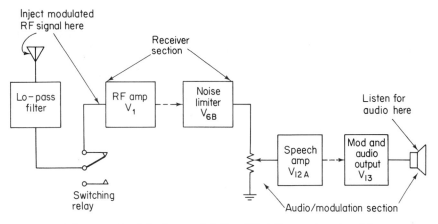

**Figure 1-12:** Injecting modulation RF at the receiver input to check receiver and audio/modulation sections

It should be noted that AVC circuit problems are often difficult to localize, because an AVC circuit uses *feedback*. For example, if the IF amplifiers are defective, the detector $V_{6A}$ and AVC circuits will not receive proper IF signals. (The AVC voltage is developed by the detector $V_{6A}$ circuit.) In turn, the lack of proper AVC voltage may cause the IF amplifiers to operate improperly. Conversely, if the AVC circuits are defective, the IF amplifiers will not receive proper AVC voltages and will not deliver a proper IF signal to the AVC circuits.

As a rule of thumb, note that if clamping the AVC line eliminates the "no-reception" symptom, the trouble is probably localized to the AVC circuits.

*If there is no transmission, or transmission is poor, but reception is normal,* check for an RF indication on the front panel meter (with the push-to-talk button pressed). If the RF indication is absent or abnormal (very low), you have localized trouble to the transmitter circuits ($V_9$, $V_{10}$, and $V_{11}$).

This symptom may also be caused by a defect in the switching control circuits. Check to see that the relay operates when the push-to-talk button is pressed, and that *all sets of contacts* switch from receive to transmit (antenna from receiver input to transmitter output, power from receiver to transmitter, audio input from receiver to microphone or audio clipper input, and the audio output to speaker is disabled). If any one of these functions is not normal, trouble is localized to the switching control circuits.

*If there is no modulation on transmission, and no* background noise in the speaker during reception (with the squelch properly set), the problem is most likely in the audio and modulation circuits. These circuits can be checked quickly by injecting an audio signal at the input (volume control or corresponding relay contacts) and listening for a tone in the speaker, as in the process for a "no-reception or reception poor" symptom, as shown in Fig. 1-11.

*If there is no modulation on transmission, but there is background noise in the speaker during reception,* the problem is likely to be in the audio clipper circuits. These circuits may be checked by injecting an audio signal at the microphone input and checking at the corresponding relay contacts, as shown in Fig. 1-13.

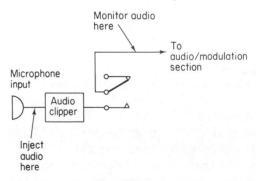

**Figure 1-13:** Checking audio signals through the audio clipper circuits

The same symptom can be produced by a *defective microphone*; this may be checked by substitution. Finally, it is possible that the symptom is caused by bad relay contacts (at the audio and modulator circuit input).

Sec. 1-8  Isolating Trouble to a Circuit                                    35

Note that some CB sets are provided with a modulation indicator. Often this indicator is a panel lamp that glows when modulation is applied (strong modulation produces a bright glow, and weak modulation produces a dull glow). Such an indicator makes it easy to localize modulation problems. For example, if the modulator lamp is not glowing (to indicate modulation), but the panel meter indicates proper RF output, the problem is between the microphone and the modulation and output circuit.

*If transmission and reception are absent or abnormal only on certain channels*, the problem is likely to be in the frequency synthesis (crystal control) circuits, or in the channel selector switching circuits. Either way, troubles producing these symptoms are usually easy to localize. However, they may not be easy to isolate once you are into the circuits, particularly on those sets where digital control is used for channel selection. Fortunately, these digital control sets are not in the majority.

*If transmission is poor but the RF indication is normal (on the front panel meter), and reception is poor (weak signal strength indication) but background noise appears normal*, the low-pass filter is a likely suspect. For example, the low-pass filter could have shorted capacitors or open coils.

The same symptoms may be caused by a *defective antenna system*. Needless to say it is possible to spend hours trying to localize problems in a perfectly good CB set if the antenna or lead-in is bad (shorted coax, improper connections, mismatching, etc.). When you are confronted with some mysterious "poor reception and poor transmission" symptom, always try the set on a known good antenna. Also, check the antenna as described in Chapter 5.

## 1-8  ISOLATING TROUBLE TO A CIRCUIT

The first two steps (symptoms and localization) of the troubleshooting procedure give you the initial symptom information about the trouble and describe the method of localizing the trouble to a *probable faulty circuit group*. Both steps involve a minimum of testing. In the isolate step, you will do extensive testing to track the trouble to a specific faulty circuit.

### 1-8.1  Isolating Trouble in IC and Plug-in Equipment

ICs are common in many solid-state CB sets. For example, in some sets the entire IF circuit or a noise blanker amplifier circuit is replaced by a single IC. All parts of the circuit group except the transformers are included in the IC. In other sets, all the components of a phase-locked loop except the crystals are contained in a replaceable module. The module may be plug-in, but it usually requires some solder connections.

In sets with ICs and replaceable modules, the trouble can be isolated to the IC or module input and output, but not to the circuits (or individual parts) within the IC. No further isolation is necessary, since parts within the IC cannot be replaced on an individual basis.

This same condition is true of some solid-state sets where groups of circuits are mounted on *sealed*, replaceable boards or cards. Note that not all modules are sealed; many have replaceable parts.

### 1-8.2  Using Diagrams in the Isolation Process

No matter what physical arrangement is used, the isolation process follows the same reasoning you have already used: the continuous narrowing down of the trouble area by making logical decisions and performing logical tests. Such a process reduces the number of tests that must be performed, thus saving time and reducing the possibility of error.

A block diagram is a convenient tool for the isolation process, since it shows circuits already arranged in circuit groups. Unfortunately, as discussed, you may or may not have a block diagram supplied with your service literature; you must work with a schematic diagram.

With either diagram, if you can recognize *circuit groups* as well as *individual circuits*, the isolation process will be much easier. For example, if you can subdivide (mentally or otherwise) the schematic diagram of the set you are servicing into circuit groups rather than individual circuits, you can isolate the group (the group from possible fault) by a single test at the input or output for the group.

The block diagram in Fig. 1-4 has been arranged into individual circuits, with each block representing a vacuum tube and its related circuit parts. You can arrange these blocks into circuit groups, as is done in Fig. 1-7. For example, the blocks representing $V_1$ through $V_6$ form the receiver circuit group. Blocks $V_7$ and $V_8$ form the frequency synthesizer, $V_9$ through $V_{11}$ form the transmitter, and $V_{12}$ and $V_{13}$ form the audio and modulation group. All CB sets will have some similar (but not identical) arrangement. Make it a practice to group the circuits mentally on the block or schematic as a first step in isolating trouble.

No matter what diagram you use, or what set arrangement is found, you are looking for three major bits of information: the *signal path* (or paths), the *signal form* (waveform, amplitude, frequency, etc.) and the *operating/adjustment controls* in the various circuits along the signal paths. If you know what signals are supposed to go where, and how the

Sec. 1-8 Isolating Trouble to a Circuit 37

signals may be affected by controls, you can isolate trouble quickly in any electronic equipment.

In Fig. 1-4, the receive signal paths are indicated by heavy dashed lines with an arrow; transmit signal paths are shown by a heavy, solid line with arrows. This arrangement is unique to the block of Fig. 1-4. Always study the service literature diagrams for any special notations. Note that both receive and transmit signal paths are shown coming from the synthesizer mixer $V_{8A}$ and going through the audio and modulation circuits $V_{12}$–$V_{13}$.

No waveforms are shown in Fig. 1-4. This is standard for most CB set diagrams (both block and schematic). The signals in a CB set are typically sinewaves (AF, IF, or RF), and the shape or form is not critical (with the possible exception of the audio section). Instead, you are interested in *amplitude* and *frequency* of the signals. Although there is no standardization, most CB service literature will show the amplitude and frequency of critical signals. Usually, this information is found in the alignment and adjustment instructions, rather than on the diagrams. About the only waveforms you will find in CB literature are those pertaining to the modulation envelope. Typically, these will be theoretical drawings, rather than duplications of actual waveforms (as is the case in TV service literature).

No operating/adjustment controls are shown in Fig. 1-4; this is typical for most CB set block diagrams. The controls are shown on the schematic diagram. Once you have arranged the individual circuits into groups, your next step is to locate all operating and adjustment controls in the group. This is just as important as locating all inputs and outputs for the circuit group.

### 1-8.3 Comparison of Signals

In its simplest form, the isolation step involves comparing the actual signals produced along the paths of the set circuits against the signals given in the service literature. This is known as *signal tracing*. The isolation step may also involve *injection* or *substitution* of signals normally found along the signal paths. Signal tracing and injection are discussed later in this chapter, as well as throughout the rest of the book. With either technique, you check and compare inputs and outputs of circuit groups and circuits in the signal paths.

In vacuum-tube sets, the input signal is injected at the grid, and the output signal is traced at the plate (or possibly the cathode). In solid-state sets, the input signal is injected at the base, and the output signal is traced at the collector (or possibly the emitter). These input/output relationships are shown in Fig. 1-14.

For a circuit group, the input is at the *first* base (or grid) in the signal path, whereas the output is at the *last* collector (or plate) in the *same* path. In any circuit group, the input signal to the group is injected at one point, and then the output signal is obtained at a point several stages further along the same signal path. To determine the signal-injection and output point of a circuit group, you must find the *first* circuit of the group in the signal path and the *final* circuit of the group in the same path.

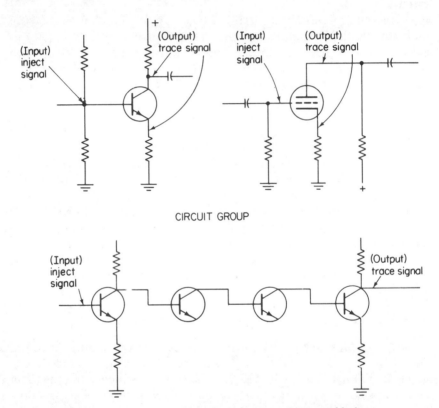

**Figure 1-14:** Input-output relationships in circuit troubleshooting

Signal paths are discussed further in Sec. 1-8.4, but now let us consider the following example. In the block diagram of Fig. 1-4, the input for the audio and modulator circuit group is at the relay contacts that switch between the audio clipper output and the noise limiter output. The output for the same circuit group is at the loudspeaker during receive and at the modulation input of the power amplifier $V_{11}$ during transmit. Note that the relay contacts are at the input of the audio group and simultaneously at the output of the receiver or clipper groups.

Sec. 1-8   Isolating Trouble to a Circuit    39

Since the relay contacts of most sets are readily identifiable (and accessible), they form good input/output test points for universal troubleshooting.

Before studying the subject of signal paths, keep the following points in mind. The symptoms and related information obtained in the previous steps (symptoms and localization) should not be discarded now or at any time during the troubleshooting procedure. From this information you will be able to identify those circuit groups which are probable trouble sources. Also, note that the physical location of the circuit groups within the set has no relation to their representation on the diagrams (block or schematic). You must consult part placement diagrams to find physical location.

### 1-8.4   Signal Paths in CB Sets

There are six basic types of signal paths, no matter what circuit group or circuit arrangement is used. These types are shown in Fig. 1-15 and are summarized as follows:

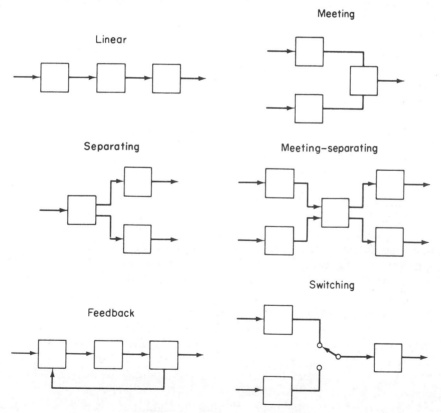

**Figure 1-15:** Types of signal paths

A *linear* path is a series of circuits arranged so that the output of one circuit feeds the input of the following circuit. Thus, the signal proceeds straight through the circuit group without any return or branch paths. As shown in the block diagram of Fig. 1-4, the paths through the IF amplifier, detectors and noise limiter, the speech amplifier and audio driver/output, as well as the RF and power amplifiers are examples of linear paths.

A *meeting* path is one in which two or more signal paths enter a circuit. The paths to the first and second mixer of the receiver group are both meeting paths.

A *separating* path is one in which two or more signal paths leave a circuit. The paths from the first synthesizer mixer could be considered separating paths.

A *meeting/separating* path is one in which a single stage has multiple inputs and multiple outputs. The paths to and from the first synthesizer mixer are true meeting/separating paths.

A *feedback* path is a signal path from one circuit to a point or circuit preceding it in the signal flow sequence. The AVC circuit from the detector back to the RF and IF amplifier stages is an example of a feedback path.

A *switching* path contains a selector of some sort (usually a switch) that provides a different signal path for each switch position. The paths in the transmit-receive relay and the channel selector circuit are switching paths.

### 1-8.5 Signal Tracing versus Signal Substitution

Both signal tracing and signal substitution (or signal-injection) techniques are used frequently in troubleshooting all types of CB sets. The choice between tracing and substitution depends on the test equipment used and the circuits involved. Signal tracing is generally used for the oscillator, frequency synthesizer, and transmitter RF amplifier circuits, because these circuits generate or amplify signals that are readily traced and need not be substituted. Signal substitution is generally used for the receiver RF, IF, noise, squelch, and audio circuits. Some signal generators designed for CB set service have outputs that simulate signals found in all major signal paths of the receiver circuits. For such a generator, signal injection is the logical choice, since you can test all the circuit groups individually (independently of other circuit groups), from antenna to the loudspeaker and/or S/RF meter. However, it is possible to troubleshoot the receiver circuits of CB sets with signal tracing alone (and this technique is recommended by many technicians).

*Signal tracing* is done by examining the signals at test points with a

Sec. 1-8  Isolating Trouble to a Circuit    41

monitoring device such as a frequency counter, oscilloscope, multimeter, or loudspeaker. In signal tracing, the input probe of the indicating or monitoring device used to trace the signal is moved from point to point, with a signal applied at a fixed point. The applied signal may be generated from an external device, or the normal signal associated with the equipment may be used (such as using the transmission from a nearby CB set).

*Signal substitution* is done by injecting an artificial signal (from a signal generator) into a circuit or circuit group (or to the complete CB set) to check performance. In signal injection, the injected signal is moved from point to point, with an indicating or monitoring device remaining fixed at one point. The monitoring can be done with external test equipment or with the set's S/RF meter and loudspeaker.

Both signal tracing and substitution are often used simultaneously in troubleshooting CB sets. For example, in dealing with the receiver circuits it is common practice to inject a modulated RF signal at the antenna and monitor the output with a meter or scope.

### 1-8.6  Half-split Technique

The *half-split technique* is based on the idea of using any test that will eliminate the maximum number of circuit groups or circuits simultaneously. This will save both time and effort. The half-split technique is used primarily in isolating trouble in a linear signal path, but it can also be used with other types of signal paths. In this system, brackets are placed at good-input and bad-output points in the normal manner, and the symptoms are studied. Unless the symptoms point definitely to one circuit or circuit group that might be the trouble source, the most logical place to make the first test is at a *convenient* test point *halfway between* the brackets.

*Example of the half-split technique.* The block diagram in Fig. 1-16 is a simplified version of the *receiver circuit group* for the CB set shown in Fig. 1-4. Note that in Fig. 1-16 the first and second mixers are combined into one block, as are the first and second IF amplifiers. Figure 1-16 thus shows the linear signal path of the received signal from the antenna to the volume control. The brackets placed at the antenna (good-input) and volume control (bad-output) show the trouble being localized to the receiver circuit group. Brackets should be placed at these points as a result of a "no reception on any channel, transmission normal" trouble symptom or a "reception poor, transmission good" symptom.

The next phase of troubleshooting is to isolate the trouble to one of the circuit groups (mixers or IF amplifiers) or to one of the individual

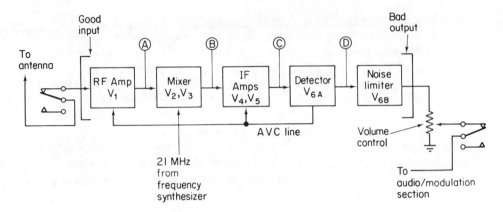

**Figure 1-16:** Simplified servicing block diagram of receiver section illustrating linear signal path from antenna to volume control

circuits (RF amplifier, detector, or noise limiter) in the linear signal path. The selection of test points during this phase depends on their *accessibility* and the *method* of troubleshooting (signal tracing or signal injection).

Assuming that test points A, B, C, and D are equally accessible (and that there are no special symptoms that would point to a particular circuit or group), test point C is the most logical point for the first test if signal tracing is used. Test point B is the next most logical choice. When using signal injection, however, test points C and D are the most logical choices. The discussion that follows describes the reasoning for making these choices.

*Half-split technique using signal tracing.* If signal tracing is used, an RF signal (at the selected channel frequency) modulated by an AF tone is introduced at the antenna. A monitoring device is then connected to monitor the signal at various test points, as shown in Fig. 1-17. The monitoring device may be a meter or oscilloscope with suitable probes or a frequency counter. The meter or scope will show the signal amplitude, but not the frequency. The counter shows the signal frequency, but not amplitude. Test equipment is discussed further in Chapters 2 and 4.

*Test point C is a logical choice for a first test.* If C is chosen first and the monitoring display is normal, you have cleared four circuits (RF amplifier, mixer, local oscillator, and IF amplifiers). You have also established the fact that there is a 21 MHz signal from the frequency synthesizer. However, there may still be defective circuits (detector and noise limiter). Note that this process divides the circuits into two groups (known good and possibly bad).

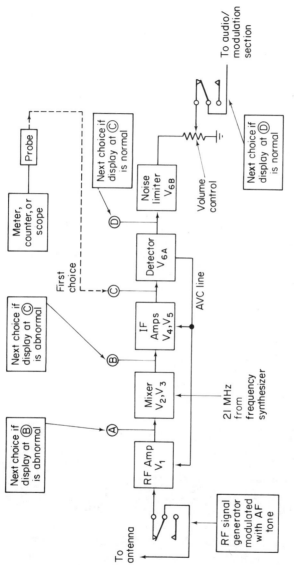

**Figure 1-17** Example of half-split technique using signal tracing

Now assume that the indication at test point C is abnormal (totally absent, low in amplitude, off-frequency etc.). The bad-output bracket can be moved to test point C, with the good-input bracket remaining at the antenna. The next logical test point is B, because B is near the *halfway point* between the two brackets. (If you choose point A instead of point B, you will confirm or deny the possibility of trouble in the RF amplifier *only*).

If the monitoring indication is abnormal at test point B, the trouble is isolated to the RF amplifier or the mixers. (The term *mixer* used here includes the local oscillator and the 21 MHz signal from the frequency synthesizer.) Additional observation at test point A will further isolate the trouble to either circuit (RF amplifier or mixers).

The final step in this half-split process is to monitor the signal at test point A. If there is an abnormal indication at A, the bad-output bracket can be moved to A and the trouble isolated to the RF amplifier. If there is a normal display at A, the good-input bracket can be moved to A and the trouble isolated to the mixers (including local oscillator).

Now, let us see what happens when a test point *other than* C is monitored first, using signal tracing.

If you choose test point A for the first test and get a normal indication, the trouble is located somewhere between A and the volume control. This means that the trouble could be in the mixers, IF amplifiers, detector, or noise limiter; you still have many test points to check.

On the other hand, if you get an abnormal signal at A, the trouble is immediately isolated to the RF amplifier. All other circuits are eliminated, but this would be a *lucky accident*, not good troubleshooting. To be performed efficiently and rapidly, the troubleshooting procedures should be based on a systematic, logical process, not on chance or luck. You will probably have as many unlucky accidents as lucky ones throughout your troubleshooting career.

The same condition holds true if test point D is chosen first (if you use signal tracing). A normal signal at D clears all circuits except the noise limiter. An abnormal signal at D still leaves the possibility of trouble in many circuits.

If test point B is chosen first and the signal is normal, this will clear three circuits (RF amplifier, mixers, local oscillator), but will leave four circuits possibly defective (both IF amplifiers, detector, and noise limiter). The opposite results will be obtained if the signal is abnormal at B.

Actually, test point B is not a bad choice for a first test using signal tracing. If test point B is more readily accessible than test point C, use B.

*Half-split technique using signal injection.* If signal injection is used, signals of the right sort (proper frequency and amplitude) are injected at test points A, B, C, D, and the antenna, as shown in Fig. 1-18.

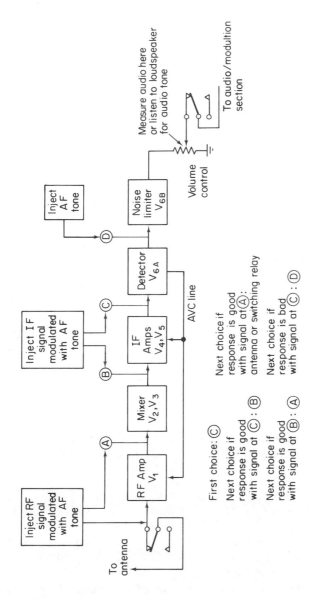

**Figure 1-18:** Example of half-split technique using signal injection

Receiver circuit response is noted at the volume control or at the loudspeaker. In this case, the signal for test points A and the antenna is at the RF frequency of the selected channel, modulated by an AF tone. The signals for B and C are at the IF frequency, also modulated by an AF tone. The signal for D is an AF tone. Note that signal injection requires several different types of signal sources at different frequencies, whereas signal tracing requires only one signal at the input.

Using signal injection with the half-split technique, the first signal is again injected at test point C. Now, however, a normal response at the volume control loudspeaker will clear the final circuits (detector and noise limiter). Under these circumstances, the next logical points for signal injection are B and A, in that order.

An absent or abnormal response at the volume control or loudspeaker (with a signal injected at C) isolates the trouble to the detector or noise limiter. Under these circumstances, the next logical test point is D.

A normal response at the volume control or loudspeaker with a signal at D, but not with a signal at C, isolates the trouble to the detector. An absent or abnormal response with a signal at D isolates the trouble to the noise limiter.

Keep in mind that these examples using the half-split technique, signal tracing, signal injection, and bracketing to isolate the trouble to a circuit group by no means cover all the possibilities that may occur. They simply illustrate the basic concepts involved when following the systematic, logical troubleshooting procedure in servicing CB sets.

### 1-8.7 Isolating Trouble to a Circuit within a Circuit Group

Once trouble is definitely isolated to a faulty circuit group, the next step is to isolate the trouble to the faulty circuit within the group. Bracketing, half-splitting, signal tracing, signal injection, and a *knowledge of the signal path* in the circuit group are all important in this step and are essentially the same methods used for isolating trouble to the circuit group.

## 1-9  LOCATING A SPECIFIC TROUBLE

The ability to recognize symptoms and to verify them with test equipment will help you to make logical decisions regarding the selection and localization of the faulty circuit group. You will also be able to isolate trouble to a faulty circuit. The final step of troubleshooting—locating the *specific* trouble—requires testing of the various branches of the faulty circuit to find the defective part.

The proper performance of the locate step will enable you to find the cause of trouble, repair it, and return the set to normal operation. As a

Sec. 1-9  Locating a Specific Trouble            47

follow-up, record the trouble so that, from the history of the set, future troubles may be easier to locate. Also, such a history may point out consistent failures that might be caused by a design error.

### 1-9.1 Locating Troubles in Replaceable Modules

Because the trend in modern CB sets (and most other electronic equipment) is toward IC and replaceable-module (PC board) design, technicians often assume it is not necessary to locate specific troubles to individual parts, that is, that all troubles can be repaired by replacement of IC packages or PC boards. Some technicians are even trained that way. The assumption is erroneous.

Although the use of replaceable modules often minimizes the number of steps required in troubleshooting, it is still necessary to check branches to parts outside the module. Front-panel operating controls are a good example of this, since such controls are not located in IC packages or on PC boards (usually); rather, they are connected to the terminal of an IC, circuit board, or other module.

### 1-9.2 Inspection Using the Senses

After the trouble is isolated to a particular circuit, the next step is to conduct a preliminary inspection using your physical senses. For example, burned or charred resistors may be observed visually or by smell. The same holds true for oil-filled or wax-filled parts such as some capacitors, coils, and transformers.

Overheated parts, such as hot transistor cases, may be located quickly by touch. The sense of hearing can be used to listen for arcing between wires or between wires and the chassis, for "cooking" or overloaded or overheated transformers, or for hum or lack of hum. Although all the senses are involved, the procedure is usually referred to as a *visual* inspection.

### 1-9.3 Testing to Locate a Faulty Part

*Testing vacuum tubes.* Vacuum tubes are relatively easy to replace, compared to transistors or IC units. For that reason, two approaches have become common as the first step in troubleshooting vacuum-tube sets: either all the vacuum tubes are tested by substitution, or they are all removed and tested on a tube tester. Neither practice is valid.

In *tube substitution*, the tubes are replaced one at a time until the set again works normally; then the *last* tube replaced is discarded, and all the other original tubes are reinserted in their respective sockets. However, this approach presents several problems.

Some oscillator, frequency synthesizer, or RF amplifier circuits may operate with one new tube and not with another because of the differences in interelectrode capacitance between the tube elements (a good tube may react like a bad tube). When removing or inserting the tubes, rocking or rotating them may result in bent pins or broken weld wires where the pins enter the envelope. If there is more than one bad tube in the set, substituting good tubes one at a time and reinserting the original tube before substituting the next tube will not locate the defective tube. Finally, if the replacement tube becomes defective immediately after substitution, there definitely is circuit trouble, and further troubleshooting is required.

*Testing all tubes on a tube tester*, as the first troubleshooting step, is also not recommended. Because this procedure has been followed religiously in the past, the practice has led to the misconception that defective tubes are the cause of all or most set failures. Even if defective tubes cause 50% (or more) of all set failures, the process of removing tubes, checking them on a sometimes marginal tube tester, and replacing them with new tubes, *as a first step without further circuit checking*, is a waste of time (and is poor troubleshooting practice).

This does not mean that a tube should never be checked first, after the trouble is isolated to a circuit. For example the power can be turned on and the tube filaments checked first for proper warm-up. If the tube envelopes are glass, a visual inspection will show whether the filament is burned out. For metal-envelope tubes, feeling the envelope will indicate if the filament is lit.

This type of test may speed the troubleshooting effort by quickly locating a tube with a burned-out filament. If the tube does not warm up properly, remove it and check it on a tube tester or substitute a new (known good) tube, whichever is most convenient. In either case, the complete circuit should be checked to determine if the tube burned out naturally because of long use or some trouble in the circuit. Simply replacing the tube without checking the rest of the circuit does not complete the process of locating the trouble. You still must determine whether the burned-out tube is the cause or the effect of the trouble.

The tube-checking procedure just described works well when the tube filaments are all connected in *parallel*, which is the usual case for vacuum-tube CB sets. With parallel filaments, when the filament of one tube burns out, that tube, and *only* that tube, will show a bad (unlit) filament. The filaments of tubes connected in *series* present more of a problem. When one filament burns out, all the filaments in the string will be unlit. This condition is shown in Fig. 1-19 and, although not typical, can be found in some CB sets. In such circuits it is more difficult to determine which filament in the string is bad.

Removing the tubes one at a time and checking their filaments for

Sec. 1-9  Locating a Specific Trouble                                49

continuity with an ohmmeter is time-consuming, and, unless handled with care, the tube may be damaged during the test, since the current from the ohmmeter (set on its lowest scale) may be high enough to burn out the filament. A better test is to measure the voltage across the filament terminals of the tube socket, provided that the bottom of the socket is accessible and all the tubes are left in their sockets. All good filaments in the string will show zero voltage, but the one that is defective (burned out) will have full voltage, which is applied to the filament string as shown in Fig. 1-19.

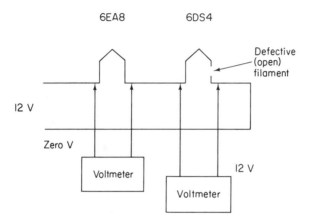

**Figure 1-19:** Method of locating defective (open) filament in vacuum-tube CB sets (with filaments connected in series)

*Testing solid-state and IC sets.* Unlike vacuum tubes, most transistors, ICs, and solid-state diodes are not easily replaced. Thus, the old electronic troubleshooting procedure of replacing tubes at the first sign of trouble is not applicable to solid-state CB sets. Instead, solid-state circuits are analyzed by testing to locate faulty parts.

*Testing the active device.* For service purposes, the vacuum tube, transistor, IC, and solid-state diode may be considered the *active devices* (or common denominator) in any electronic circuit. Because of their key position in the circuit, these devices are useful in evaluating the operation of the entire circuit (through signal, voltage, and resistance tests). Preliminary tests at the terminals of the active device will usually locate the trouble quickly.

*Signal testing.* Usually, the first step in circuit-testing is to analyze the *output signal* of the circuit or the output signal of the active device. (This output is generally at the plate of a vacuum tube or at the collector of a transistor.) Of course, in some circuits (such as power supplies) there is no output signal as such, and some circuits show no signal of any significance.

In addition to checking for the presence of a signal, the signal must be analyzed in detail for proper amplitude and frequency. As discussed throughout this book, a careful analysis of signals will often pinpoint the branch of a circuit most likely to be defective.

*Transistor and diode testers.* It is possible to test transistors and diodes in circuit, using in-circuit testers (such testers are discussed further in Chapter 2). These devices are usually adequate for transistors used at lower frequencies, particularly in the audio range. However, most in-circuit tests will not show the high-frequency characteristics of transistors. (The same is true for out-of-circuit transistor and diode testers.) For example, it is possible for a transistor to perform well in the audio section, but be hopelessly inadequate in the RF, IF oscillator, or frequency synthesizer sections of the set.

*Voltage testing.* After signal analysis and/or in-circuit tests, the next logical step is voltage measurement at the active device terminals or leads. Always pay particular attention to those terminals which show an abnormal signal, since they are most likely to show an abnormal voltage (but not always).

When properly prepared service literature is available (with signal, voltage, and resistance information), the actual voltage measurements may be compared with the normal voltages listed in the service literature. This test will often help isolate the trouble to a single branch of the circuit.

*Relative voltages in solid-state sets.* It is often necessary to troubleshoot solid-state circuits without benefit of adequate voltage and resistance information. This can be done by using the schematic diagram to make a logical analysis of the relative voltages at the transistor terminals. For example, in an NPN transistor the base must be positive in relation to the emitter if there is to be emitter-collector current flow. That is, the emitter-base junction will forward-biased when the base is more positive (or less negative) than the emitter. The problem of troubleshooting solid-state CB sets on the basis of relative voltages is further discussed in Chapter 4.

*Resistance measurements.* After signal and voltage measurements are taken, it is often helpful to make resistance measurements at the same point on the active device where an abnormal signal and/or voltage is

Sec. 1-9 Locating a Specific Trouble 51

found. Of course, other points (for example, suspected parts) in the circuit may also be checked by resistance measurement, or a continuity check can be made to find point-to-point resistance of the suspected branch. Considerable care must be used when making resistance measurements in solid-state circuits, since the junctions of transistors act like diodes, and when biased with the right polarity (by the ohmmeter battery) the diodes will conduct and produce false resistance readings. This condition is discussed in Sec. 1-9.8 and in Chapter 4.

*Current measurements.* In rare cases, the current in a particular circuit branch can be measured directly with an ammeter. However, it is usually simpler and more practical to measure the voltage and resistance of a circuit and then calculate the current.

### 1-9.4 Signal Measurements

When testing to locate trouble, the signals are measured with the circuit in operation and usually with an input signal (or signals) applied. The signals can originate at a test generator or from another CB set, whichever is convenient for the particular measurement. The signal tracing and signal injection techniques described in Sec. 1-8.6 are typical. If you use signal information found in the service literature, follow all the notes and precautions described in the literature. Usually, the literature will specify the positions of operating controls, typical input signal amplitudes, and so on.

Note that there is a relationship between signals and trouble symptoms. Complete failure of a circuit will usually result in the absence of a signal, whereas a poorly performing circuit will usually produce an abnormal signal. For example, an improperly adjusted circuit could produce an off-frequency signal, or a defective circuit could produce a low-amplitude signal.

### 1-9.5 Voltage Measurements

In order to locate trouble, voltages are measured with the circuit in operation but usually with no signals applied. In some cases, the voltage measurements can be made with the set operating. If you are using the voltage information found in service literature, follow all the notes and precautions. Usually the service literature will specify the position of operating controls, typical input voltages, and so on.

In most CB set literature the voltages are given in chart form, as shown in Fig. 1-20. Occasionally, voltages will be found on the schematic diagram. In either system, you must find the physical location of the terminals where the voltages are to be measured. Occasionally, CB service literature will follow the military style and show the actual physical location of the terminals.

Because of the safety practice of setting a voltmeter to its highest scale before making measurements, the terminals having the highest

SQUELCH AMPLIFIER DC VOLTAGE CHART

| Transistor | Mode | DC Voltage In Volts | | |
|---|---|---|---|---|
| | | Emitter | Base | Collector |
| $Q_8$ | squelch | 10.0 | 8.8 | 9.5 |
| | unsquelch | 4.4 | 8.0 | 0 |
| $Q_9$ | squelch | 1.6 | 2.2 | 1.6 |
| | unsquelch | 0 | 0 | 3.5 |

**Figure 1-20:** Example of voltage information found in CB service literature

voltage (vacuum-tube plate or transistor collector) should be checked first. (In some solid-state sets, the collector is grounded, and the emitter has the highest voltage.) Then the elements having lesser voltage should be checked in descending order.

If you have had any practical experience in troubleshooting, you know that voltage (as well as resistance and signal) measurements are seldom identical to those listed in the service literature. This brings up an important question concerning voltage measurements: "How close is good enough?" In answering this question, several factors must be considered.

The tolerances of the resistors, which greatly affect the voltage readings in a circuit, may be 20, 10, or 5%. Resistors with 1% (or better) tolerances are used in some critical circuits. The tolerances marked or color-coded on the parts are thus one important factor. The wide range of characteristics of transistors and diodes will cause variations in voltage readings.

The accuracy of test instruments must also be considered. Most voltmeters have accuracies of a few percent (typically 5 to 10%). Precision laboratory meters (generally not used in CB troubleshooting) have a much greater accuracy.

In order to operate properly, the voltages of critical circuits must fall within a narrow range of tolerances (at least 10% and probably closer to 3%). However, many circuits will operate satisfactorily within tolerance ranges of 20 to 30%.

Sec. 1-9 Locating a Specific Trouble    53

Generally, the most important factors to consider in voltage measurement accuracy are the symptoms and the output signal. If no output signal is produced by the circuit, you should expect a wide variation of voltages in the trouble area. Trouble that results in circuit performance that is just out of tolerance range may cause only a slight change in circuit voltages.

### 1-9.6 Resistance Measurements

Resistance measurements must be made with no power applied. However, in some cases various operating controls must be in certain positions to produce resistance readings similar to those found in the service literature. This is particularly true of controls that have variable resistances.

Always observe any notes or precautions found in the service literature. In any circuit, always check that the filter capacitors are discharged before making resistance measurements. After all safety precautions and notes have been observed, measure the resistance from the terminals of the active device to the chassis (or ground) or between any two points that are connected by wiring or parts.

In most CB service literature, resistance information is given in chart form, as shown in Fig. 1-21. Sometimes resistance information will be found on the schematic diagram. Do not be surprised if you find CB service literature with little or no resistance data. Often, the only resis-

SQUELCH AMPLIFIER RESISTANCE CHART

| Transistor | Mode | Resistance In Ohms | | |
|---|---|---|---|---|
| | | Emitter | Base | Collector |
| $Q_8$ | squelch | 3.9 K | 38 K | 10.9 K |
| | unsquelch | 500 | 38 K | 0 |
| $Q_9$ | squelch | 1 K | 2.7 K | 10.9 K |
| | unsquelch | 1 K | 2.7 K | 10.9 K |

**Figure 1-21:** Example of resistance information found in CB service literature

tances given are the values of resistors and the dc resistance of coils and transformers. In well-prepared literature, you will find the resistance from all terminals of the active device.

There is good reason for the omission of active device resistances: if there is a condition in any active device terminal circuit that will produce an abnormal resistance (say an open or shorted resistor or a resistor that has changed drastically in value), the voltage at that terminal will be abnormal. If such an abnormal voltage reading is found, it is then necessary to check out each resistance in the terminal circuit on an individual basis.

Because of the *shunting effect* of other parts connected in parallel, the resistance of an individual part or circuit may be difficult to check. In such cases, it is necessary to disconnect one terminal of the part being tested from the rest of the circuit. This will leave the part open at one end, and the value of resistance measured is of that part only.

Keep in mind that when making resistance checks that a zero reading indicates a short circuit, and an infinite reading indicates an open circuit. Also remember the effect of the transistor junctions (acting as a forward-biased diode when biased on). The problems of resistance measurements are discussed further in Sec. 1-9.8 and in Chapter 4.

### 1-9.7 Duplicating Signal, Voltage, and Resistance Measurements

If you are responsible for service of a particular type or model of CB set, it is strongly recommended that you duplicate all of the signal, voltage, and resistance measurements found in the service literature with test equipment of your own. This should be done with a known good set that is operating properly. Then when you make measurements during troubleshooting you can spot even slight variations in voltage, signal amplitude, etc. Always make the initial measurements with test equipment that would normally be used during troubleshooting. If more than one set of test equipment is used, make the initial measurements with all available test equipment and record the variations.

### 1-9.8 Using Schematic Diagrams

Regardless of the type of trouble symptom, the actual fault can be traced eventually to one or more of the circuit parts (vacuum tubes, transistors, ICs, diodes, resistors, capacitors, coils, transformers, and so forth). The checks of signals, voltage, and resistance will then indicate which branch within a circuit is at fault. Finally, you must locate the particular part that is causing the trouble in the branch.

In order to do this, you must be able to read a schematic diagram. These diagrams show what is inside the blocks on the block diagram and provide the final picture of the CB set. Often, you must service CB sets with the aid of nothing more than a schematic diagram. If you are fortunate, the diagram may also show some voltages.

Sec. 1-9   Locating a Specific Trouble                                          55

*Examples of using schematic diagrams.* Figure 1-16 shows the block diagram for the receiver circuits of a CB set. Below are some examples of how the schematic diagrams for similar circuits can be used in troubleshooting. Although these examples show only portions of the receiver circuits (and are shown as solid-state circuits instead of vacuum-tube), the same basic troubleshooting principles apply to all of the circuits in a CB set. In order to follow these examples, you must be able to use basic test equipment properly. If you are unsure of how such test equipment is used in CB troubleshooting, read Chapters 2 and 4 before you study these examples. (You may also read Chapters 2 and 4 in conjunction with these examples.)

*Example 1.* Assume that the receiver circuits of Fig. 1-16 are being serviced, using signal injection as shown in Fig. 1-18. The trouble was initially isolated to the receiver circuits by a "no reception, good transmission" trouble symptom. An IF signal modulated by an AF tone is injected at test point B, and a proper response is noted at the volume control (receiver circuit output). The 21 MHz signal from the frequency synthesizer to the mixer is normal (on-frequency and of correct amplitude). However, there is no response when an RF signal (at the selected channel frequency) modulated by an AF tone is applied at the mixer input (test point A). Now you have localized trouble to the mixer circuit.

The next step is to measure the voltages at the terminals of the active device in the mixer. These are shown as the collector, emitter, and base of $Q_2$ in Fig. 1-22. If any of the $Q_2$ elements show an abnormal voltage, the

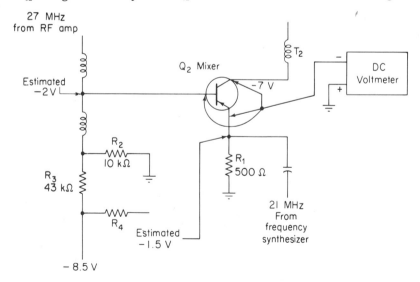

**Figure 1-22:**   Measuring voltages at elements of $Q_2$

resistance of that element should be checked first. Note that the collector voltage is specified as −7 V. Neither the base and emitter voltages nor any of the resistance values are given. This lack of information is typical for many CB sets. Thus, you must be able to interpret schematic diagrams to estimate *approximate* voltages.

For example, the voltage at the junction of $R_3$ and $R_4$ is given as −8.5 V. This is logical because the source is 9 V. The value of $R_2$ is approximately 25% of the value of $R_3$. Thus, the drop across $R_2$ is about 25% of −8.5 V, or *approximately* −2 V, and the base should be about −2 V. If $Q_2$ is silicon, the emitter will be about 0.5 V different from the base, or about −1.5 V (emitter more positive or less negative, in this case). If $Q_2$ is germanium, the base-emitter differential will be about 0.2 V, and the emitter should be approximately −1.8 V.

Keep in mind that this method of interpreting the schematic will give you *approximate* voltages only. In practical troubleshooting, the voltage *differentials* between circuit elements and transistor electrodes are the most important factor. Troubleshooting solid-state sets based on voltage differentials is discussed fully in Chapter 4.

Now assume that there is no voltage at the collector, but that the base and emitter show what appear to be normal or logical voltages. The next step is to remove power, discharge $C_{14}$ and $C_{15}$ (if necessary), and measure the collector resistance (to ground), as shown in Fig. 1-23. Because

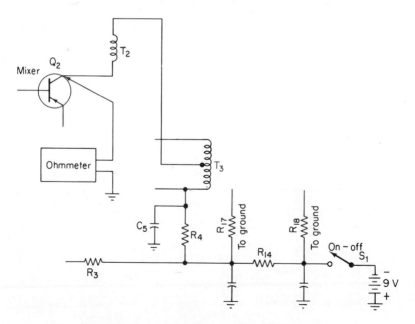

**Figure 1-23:** Measuring resistance at collector of $Q_2$

Sec. 1-9   Locating a Specific Trouble                                   57

no resistance values are given for the elements of $Q_2$, you must use the schematic to estimate the approximate values.

Of course, a zero resistance at the collector indicates a short; for example, capacitor $C_5$ could be shorted. On the other hand, an infinite resistance indicates an open circuit. For example, the coil windings of $T_2$ and $T_3$ could be open, or $R_4$ could be burned out and open. It is usually easy to locate the fault when you find such extreme resistance readings.

However, a resistance reading that falls between these two extremes does not provide a really sound basis for locating trouble. To make the problem worse, the effect of solid-state devices in the circuit can further confuse the situation. For example, assume that the fault is an open $T_3$ winding, as shown in Fig. 1-24. This will result in a no-voltage reading

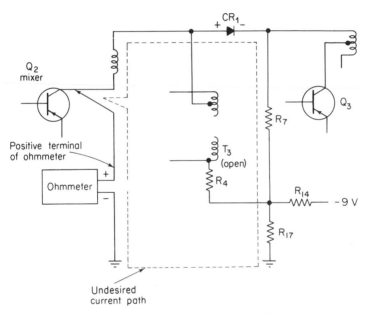

**Figure 1-24:** Undesired current path when measuring resistance at collector of $Q_2$

at the collector of $Q_2$. However, it is still possible to measure a resistance to ground from the $Q_2$ collector if the following conditions are met: Assume that the ohmmeter leads are connected so that the *positive* terminal of the ohmmeter battery is connected to the $Q_2$ collector. This will forward-bias the AVC diode $CR_1$ and cause the ohmmeter to measure the resistance across $R_7$ (the collector supply of $Q_3$) and $R_{17}$. Also, if the collector of $Q_2$ is made positive in relation to the base, the $Q_2$

base-collector junction will be forward-biased, resulting in possible current flow.

The problem illustrated in Fig. 1-24 can be eliminated by *reversing the ohmmeter leads* and measuring the resistance both ways. If there is a difference in the resistance values with the leads reversed, check the schematic for possible forward-bias conditions in diodes and transistor junctions in the associated circuit.

*Example 2.* Assume that the receiver circuits shown in Fig. 1-16 are being serviced and that the trouble is isolated to the detector. In this example, the detector is solid-state and is given as $CR_2$ rather than $V_{6A}$. Also, the noise limiter is given as $Q_6$. However, the same basic procedure is used to isolate trouble to the detector; that is, an audio signal injected at the base of $Q_6$ (test point D of Fig. 1-16) produces a good response on the receiver output, as shown in Fig. 1-25. However, no response is

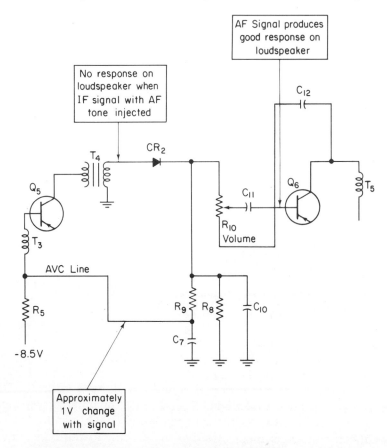

**Figure 1-25:** Example of troubleshooting detector-AVC circuits

## Sec. 1-9 Locating a Specific Trouble

obtained when an IF signal modulated by an AF tone is injected at transformer $T_4$ in Fig. 1-25. (This is test point C in Fig. 1-16.) Or, using signal tracing, an oscilloscope or meter connected across the secondary winding of $T_4$ shows that an AF modulated signal is available at the detector, but that no AF signal appears at the base of $Q_6$.

The detector $CR_2$ has two outputs: an AF signal and an AVC voltage. Under no-signal conditions (no CB signals on the channel to which the set is tuned), electrons flow through $R_9$ and $R_8$ to ground, producing approximately $-1$ V bias at the junction of $R_5$ and $R_9$. When a normal signal is received, electrons flow through $R_8$ and $CR_2$. This flow (opposite to that produced under no-signal conditions) opposes the no-signal bias and produces approximately 0 V at the junction of $R_5$ and $R_9$. This tends to reduce the gain of $Q_2$ (a PNP transistor) and thus tends to offset the signal passing through the IF stages.

The amount of AVC voltage is usually not critical, but the fact that the AVC voltage *changes* with changes in the IF signal is important. The function of the AVC circuit can be checked by measuring the voltage on the AVC line with and without a signal applied to $T_4$. If there is a change of approximately 1 V with signal, the AVC function and detector $CR_2$ may be considered normal.

With $CR_2$ established as normal but with no signal at the base of $Q_6$, volume control $R_{10}$ or coupling capacitor $C_{11}$ are logical suspects; they are probably open or have broken leads. Capacitor $C_{12}$ may also be a possible suspect, but it is not as likely as $C_{11}$ or $R_{10}$. If $C_{12}$ is shorted or leaking, this will show up in an abnormal voltage (a very large negative voltage on the AVC line). If $C_{12}$ is open, the frequency-response characteristics of the $Q_6$ circuit will be poor, but there will still be a signal at the base of $Q_6$.

The condition of $C_{11}$ may be checked in two ways (without removal from the circuit). First, as shown in Fig. 1-26, a signal (at an audio frequency) can be injected on both sides of $C_{11}$. If the signal is heard on

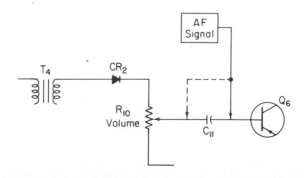

**Figure 1-26:** Localizing trouble by means of signal injection

the loudspeaker (or appears at the output of the receiver circuits) when injected at the base of $Q_6$, but is not heard when injected at the junction of $R_{10}$ and $C_{11}$, capacitor $C_{11}$ is probably defective. Second, as shown in Fig. 1-27, a signal can be injected ahead of the detector and traced on both sides of $C_{11}$ (using an oscilloscope, meter, or possibly a frequency counter). The signal should appear substantially the same on both sides of $C_{11}$. If not, $C_{11}$ is probably defective.

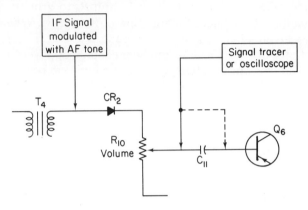

**Figure 1-27:** Localizing trouble by means of signal tracing

The condition of $R_{10}$ may also be checked by signal tracing or signal injection. However, as a final test, the power should be removed and the resistance of $R_{10}$ measured. To measure potentiometer resistance, connect the ohmmeter from the arm to one end of the winding, as shown in Fig. 1-28. Then, vary the control from one end to the other. Repeat the

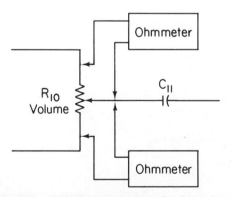

**Figure 1-28:** Measuring resistance of volume control $R_{10}$

test with the ohmmeter connected from the arm to the opposite end of the winding. The resistance indication should vary smoothly, with no jumps or flicking of the ohmmeter needle (or jitters of the display if a

Sec. 1-9 Locating a Specific Trouble 61

digital ohmmeter is used). Such jumps may mean bad spots (open or poor contact) on the potentiometer winding.

### 1-9.9 Internal Adjustments during Trouble Localization

Keep in mind that adjustment of controls (both internal-adjustment controls and operational controls) can affect circuit conditions and may lead to false conclusions during troubleshooting. For example, the amplitude of the signal at the base of $Q_6$ (Figs. 1-26 and 1-27) is set directly by the volume control $R_{10}$, which is an operational control. The amplitude of both the signal and the AVC voltage from the detector $CR_2$ can be affected by adjustment (or alignment) of the IF transformers (as well as the oscillator and the RF transformers).

If the signal at $Q_6$ is very low, it could be that the volume control is set to the minimum position. Of course, because the volume control is an operational control, a run-through of the operating sequence at the beginning of troubleshooting will pinpoint such an obvious condition. However, a low output from the detector can be caused by poor alignment of the IF transformers. Because the IF transformers require internal adjustments, poor alignment will not be detected through the use of operating procedures. This condition, or a similar one, may lead to one of two unwise courses of action.

Following the first course, you might launch into a complete alignment procedure (or whatever internal adjustments are available) once you have isolated the trouble to a circuit and are trying to locate the specific defect. No internal control, no matter how inaccessible, is left untouched! You may reason that it is easier to make adjustments than to replace parts. Such a procedure will eliminate improper adjustment as a possible fault, but it may also create more trouble than it repairs. Indiscriminate internal adjustment is the technician's equivalent of operator trouble. In the second instance, you might replace part after part when a simple screwdriver adjustment would repair the problem. This course is usually taken because of the inability to perform the adjustment procedures or a lack of knowledge concerning the control's function in the circuit. Either way, a study of the service literature should resolve the situation.

But there is a middle ground. Do not make any internal adjustments during the troubleshooting procedure until trouble has been isolated to a circuit, and then only when the trouble symptom or test results indicate possible maladjustment.

For example, assume that an oscillator is provided with an internal adjustment control that sets the frequency of oscillation. If signal measurement at the circuit output shows that the oscillator is off-frequency, it is logical to adjust the frequency control. However, if signal measurement shows a very low output from the oscillator (but on-frequency), adjustment of the frequency control during troubleshooting could cause further problems.

An exception to this rule occurs when the service literature recommends alignment or adjustment as part of the troubleshooting procedure. Generally, alignment and adjustment are checked after testing and repairs have been completed. This assures that the repair procedure (replacement of parts) has not upset circuit adjustment.

### 1-9.10 Trouble Resulting from More than One Fault

A review of all the symptoms and test information obtained thus far will help you verify that the part located as the sole trouble or isolate other faulty parts. This is true whether the malfunction of these parts is caused by the isolated part or by some entirely unrelated problem.

If the isolated malfunctioning part can produce all the normal and abnormal symptoms and indications that you have accumulated, you may logically assume that it is the sole cause of trouble. If it cannot, you must use your knowledge of electronics and the set to determine what other part or parts could have become defective and produced all the symptoms.

When one part fails, it often causes abnormal voltages or currents that could damage other parts. Trouble is often isolated to a faulty part that is a result of the original trouble, rather than to its source. For example, assume that the troubleshooting procedure thus far has isolated a transistor as the cause of trouble and that the transistor is burned out. What would cause this? Excessive current can destroy the transistor by causing internal shorts or by altering the characteristics of the semiconductor material, which is sensitive to temperature. Thus, the problem becomes a matter of finding how such excessive current was produced.

Excessive current in a transistor can be caused by an extremely large input signal, which will overdrive the transistor. Such an occurrence indicates a fault somewhere in the circuitry ahead of the transistor input. Power surges (intermittent excessive outputs from the power supply) may also cause the transistor to burn out, and are, in fact, a common cause of transistor burnout. All these conditions should be checked before a new transistor is placed into the circuit.

Some other typical malfunctions, along with their common causes, include:

1. Burned-out transistor caused by *thermal runaway*. A temporary increase in transistor current heats the transistor, causing a further increase in current, which results in more heat. This process continues until the heat-dissipation capabilities of the transistor are exceeded. Bias-stabilization circuits are gener-

Sec. 1-9  Locating a Specific Trouble    63

ally included in most well-designed transistor CB sets. Power transistors (and any transistor that dissipates more than about 1 W) are usually mounted on heat sinks to dissipate the heat.

 2. Power-supply overload caused by a short circuit in some portion of the voltage distribution network.

 3. Burned-out transistor in shunt-feed system caused by shorted blocking capacitor.

 4. Blown fuses caused by power supply surges or shorts in filtering (power) networks.

It is obviously impractical to list all the common malfunctions and their related causes that you may find in troubleshooting CB sets. Generally, when a part fails, the cause is an operating condition that exceeded the maximum ratings of the part. However, it is quite possible for a part simply to "go bad" in any CB set, no matter how well designed.

The operating condition that causes a failure can be temporary and accidental, or it can be a basic design problem, as a history of repeated failures would indicate. No matter what the cause, your job is to find the trouble, verify the source or cause, and then repair it.

### 1-9.11  Repairing Troubles

In a strict sense, repairing the trouble is not part of the troubleshooting procedure. However, repair is an important part of the total effort involved in getting the set back in operation. Repairs must be made before the set can be checked out and declared ready for operation.

Never replace a part if it fails a second time unless you make sure that the cause of trouble is eliminated. Actually, the cause of trouble should be pinpointed before you replace a part the first time. However, this is not always practical. For example, if a resistor burns out because of an intermittent short and you have cleared the short, the next step is to replace the resistor. However, the short could recur and burn out the replacement resistor. If this happens, you must recheck every element and lead in the circuit.

When replacing a defective part, an *exact replacement* should be used if it is available. If the part is in any circuit that can affect the transmitter output (frequency, output power, percentage of modulation, etc.) you *must use an exact replacement* recommended by the manufacturer. If not, the set may no longer be type accepted, and you will be in violation of FCC regulations (refer to Sec. 1-1.6).

In noncritical circuits, such as those of the receiver, if an exact replacement is not available and the original part is beyond repair, an *equivalent or better* part should be used. *Never* install a replacement part that has characteristics or ratings inferior to those of the original.

Another factor to consider during repair is that the replacement part be installed in the *same physical location* as the original, with the same lead lengths and so on, if at all possible. This precaution is optional in most low-frequency or dc circuits but must be followed for high-frequency applications. In RF, IF, oscillator, and frequency synthesizer circuits (all high-frequency circuits), changing the location of parts or the length of leads may detune the circuit, or otherwise put it out of alignment.

### 1-9.12 Operational Checkout

Even after the trouble has been found and the faulty part located and replaced, the troubleshooting effort is not necessarily completed. An operational check is necessary to verify that the set is free of all faults and is performing properly again. Never assume that simply because a defective part has been located and replaced, the set will automatically operate normally again. In practical troubleshooting any CB set, never assume anything; prove it!

Run the set through its complete operating sequence, on all channels. In this way you will make sure that one fault has not caused another. Follow the procedure found in the service literature (when available). When you are servicing a CB set for someone else, have that person go through the entire sequence (if practical), but verify operation yourself.

When the operational check is completed and the set is again operating normally, make a brief record of the symptoms, the faulty part, and the remedy. This is particularly helpful when you must service the same CB set on a regular basis or when you must troubleshoot similar sets. Even a simple record of troubleshooting will give you a valuable history of the set for future reference.

If the set does not perform properly during the operational checkout, you must continue troubleshooting. If the symptoms are the same as, or similar to, the original trouble symptoms, retrace your steps, one at a time. If the symptoms are entirely different, you may have to repeat the entire troubleshooting procedure from the beginning. However, this is usually not necessary. For example, assume, that the receiver circuits do not check out because a replacement IF amplifier transistor has detuned the circuit. In this case, you should repair the trouble by IF alignment rather than by returning to the first troubleshooting step and repeating the entire procedure. Keep in mind that you have arrived at the defective circuit or part by a systematic procedure. Thus, retracing your steps—one at a time—is the logical course of action.

# 2
# CB SERVICE EQUIPMENT

The test equipment used in CB service is basically the same as that used in other fields of electronics. That is, most service procedures are performed using meters, signal generators, frequency counters, oscilloscopes, power supplies, and assorted clips, patchcords, etc. Theoretically, all CB service procedures could be performed using conventional test equipment, provided that the oscilloscopes had the necessary gain and band-pass characteristics, the frequency counters covered the necessary range, the signal generators covered the appropriate frequencies, and so on. However, there are some specialized test instruments that greatly simplify CB service, such as SWR meters, field strength meters, RF wattmeters, and dummy loads. Also, there are specialized versions of basic test equipment that have been developed specifically to simplify CB service. We shall concentrate on this specialized equipment throughout this chapter.

It is not the purpose of this chapter to promote one type of test equipment over another (and one manufacturer over another). Instead, the chapter is devoted to the basic operating principles of test equipment types in common use. Each type of test equipment is discussed in turn. You may then select the type of equipment best suited to your own needs and pocketbook.

Although complicated theory has been avoided, the following discussions cover the way each type of equipment is used in CB service, and what signals or characteristics are to be expected from each type of equipment. They also describe how the features and outputs found on present-day test equipment relate to specific problems in CB service.

A thorough study of this chapter will familiarize you with the basic principles and operating procedures for typical equipment used in general CB service. It is assumed that you will take the time to become equally familiar with the principles and operating controls for any particular test equipment you use. Such information is contained in the service literature for the particular equipment.

It is absolutely essential that you become thoroughly familiar with your own particular test instruments. No amount of textbook instruction will make you an expert in operating test equipment; it takes actual practice.

It is strongly recommended that you establish a routine operating procedure, or sequence of operation, for each item of service equipment. This approach will save time and familiarize you with the capabilities and limitation of your own equipment, thus minimizing the possibility of false conclusions based on unknown operating conditions.

## 2-1 SAFETY PRECAUTIONS IN CB SERVICE

In addition to a routine operating procedure, certain precautions must be observed during operation of any electronic test equipment during service. Many of these precautions are the same as for all types of test equipment; others are unique to special test instruments such as meters, oscilloscopes, and signal generators. Some of the precautions are designed to prevent damage to the test equipment or to the circuit where the service operation is being performed. Others are meant to prevent injury to you. Where applicable, special safety precautions are included throughout this book.

The following general safety precautions should be studied thoroughly and then compared to any specific precautions called for in the test equipment service literature and in the related chapters of this book.

1. *The most hazardous element in CB service is the antenna of a base station.* Each year, thousands of base station antennas make accidental contact with nearby high voltage lines, resulting in damage, injury and death! If you are responsible for installing a base-station antenna, make certain that the antenna cannot possibly come in contact with high voltage lines. Remember that the lines can fall across the antenna or lead-in, or, more likely, the antenna or lead-in may fall across the line. Either way, the situation is highly dangerous! If you must service any base station or fixed antenna, perhaps to make SWR or field-strength tests, always check for the presence of high-voltage lines before con-

## Sec. 2-1  Safety Precautions In CB Service

necting any test equipment to the antenna lead-in. Keep in mind that if the antenna is in contact with any high voltage, the voltage will be transferred through the lead-in and test equipment to you!

2. The next most hazardous problem in CB service arises from a set that operates from 115 V line power. This voltage is sufficient to cause death. If the set uses vacuum tubes, there will be voltages in excess of 115 V, typically 300 V. These voltages are even more dangerous. Fortunately, most present-day sets are solid-state and operate at about 12 V. Some exceptions are the base-station sets that convert 115 V line power to 12 V, or the sets that use external converters to change the 115 V ac into 12 V dc.

3. Many service instruments are housed in metal cases. These cases are connected to ground of the internal circuit. For proper operation, the ground terminal of the instrument should always be connected to the ground of the set being serviced. Make certain that the chassis of the set being serviced is not connected to either side of the ac line or to any potential above ground. If there is any doubt, connect the set being serviced to the power line through an *isolation transformer*.

4. Remember that there is always danger in servicing sets that operate at hazardous voltages (such as the higher voltages used by vacuum-tube sets), especially as you pull off the set cover and reconnect the power cord! Always make some effort to familiarize yourself with the set before servicing it, bearing in mind that high voltages may appear at *unexpected points* in a defective set.

5. It is good practice to remove power before connecting test leads to high-voltage points. It is preferable to make all service connections with the power removed. If this is impractical, be especially careful to avoid accidental contact with circuits and objects that are grounded. Keep in mind that even low-voltage circuits may present a problem. For example, a screwdriver dropped across a 12 V line in a solid-state set can cause enough current to burn out a major portion of the set, possibly beyond repair. Of course, this problem is nothing compared to the possibility of injury to yourself! Working with one hand away from the set and standing on a properly insulated floor lessens the danger of electrical shock.

6. Capacitors may store a charge large enough to be hazardous. Discharge filter capacitors before attaching test leads.

7. Remember that leads with broken insulation offer the additional hazard of high voltages appearing at exposed points along the leads. Check test leads for frayed or broken insulation before working with them.

8. To lessen the danger of accidental shock, disconnect test leads immediately after the test is completed.

9. Remember that the risk of severe shock is only one of the possible hazards. Even a minor shock can place you in danger of more serious risks, such as a bad fall or contact with a higher voltage source.

10. The experienced service technician is continuously on guard against injury and does not work on hazardous circuits unless another person is available to assist in case of accident.

11. Even if you have considerable experience with the test equipment used in service, always study the service literature of any instrument with which you are not thoroughly familiar.

12. Use only shielded leads and probes. Never allow your fingers to slip down to the meter probe tip when the probe is in contact with a "hot" circuit.

13. Avoid vibration and mechanical shock. Most electronic test equipment is delicate.

14. Study the circuit being serviced before making any test connections. Try to match the capabilities of the instrument to the circuit being serviced.

## 2-2 SIGNAL GENERATORS

The signal generator is an indispensible tool for practical CB service. Without it you are entirely dependent on signals transmitted by another CB set, and you are limited to signal tracing only. This means that you have no control over frequency, amplitude, or modulation of such signals and have no means for signal injection. With a signal generator of the appropriate type, you can duplicate transmitted signals or produce special signals required for alignment and test of all circuits found in a CB set. Also, the frequency, amplitude, and modulation characteristics of the signals can be controlled so that you can check operation of the receiver circuits under various signal conditions (weak, strong, normal, or abnormal signals).

### 2-2.1 Signal Generator Basics

An oscillator (audio, RF, pulse, etc.) is the simplest form of signal generator. At the most elementry level of troubleshooting, a single-stage AF or RF oscillator can serve the purpose of providing a signal source. The special test sets described in Sec. 2-13 generally include such basic oscillator circuits. Beyond the simplest troubleshooting, most comprehensive CB service requires an RF and an AF generator, and possibly a pulse generator. Another instrument that may be useful in CB work is the probe-type (or pencil-type) generator.

### 2-2.2 Probe or Pencil Generators

These generators (also known as pencil-type noise generators, signal injectors, and by various other names) are essentially solid-state pulse generators or oscillators with a fast-rise waveform output and no adjustments. The fast-rise output produces simultaneous signals over a wide frequency range. The output signals may be used to troubleshoot the receiver and audio/modulation circuits of the CB set. However, except in basic troubleshooting situations, such an instrument has many obvious drawbacks.

For example, to check the selectivity of the receiver circuits, the signal source must be variable in amplitude. To check the detector or

Sec. 2-2 Signal Generators 69

audio portions of the receiver circuits, the signal source must be capable of internal and/or external modulation. These characteristics are not available in the pencil-type unit. As a result, even the least expensive shop-type (or even kit-type) generators have many advantages over the pencil generators.

### 2-2.3 RF Signal Generators

There are no basic differences between shop-type and lab-type generators; that is, both instruments will produce RF signals capable of being varied in frequency and amplitude, and capable of internal and external modulation. However, the laboratory-type instruments incorporate several refinements not found in shop equipment, as well as a number of quality features (this accounts for the wide difference in price). Following is a summary of the differences between shop and lab RF generators.

*Output meter.* In most shop generators, the amplitude of the RF output is either unknown or approximated by means of dial markings. The lab generator incorporates an output meter. This meter is usually calibrated in microvolts so that the actual RF output may be read directly. If you use a shop generator without a built-in output meter, you must monitor the output with an *external meter*. Keep in mind that the meter must be capable of indicating output signals in the order of a few microvolts ($\mu$V) to properly perform all receiver circuit checks of some CB sets. Meters are discussed in Sec. 2-4.

*Percentage-of-modulation meter.* Most shop generators have a fixed percentage of modulation (usually about 30%). Lab generators provide for a variable percentage of modulation and a meter to indicate this percentage. Some generators have two meters (one for output amplitude and one for modulation percentage). Some generators use the same meter for both functions.

*Output uniformity.* Shop generators vary in output amplitude from band to band. Also, shop generators usually cover their entire frequency range by means of harmonics. Lab generators have a more uniform output over their entire operating range and cover the range with pure fundamental signals.

*Wideband modulation.* Generally, the oscillator of a shop generator is modulated directly. This can result in undesired frequency modulation. The oscillator of a lab generator is never modulated directly (unless it is designed to produce an FM output); instead, the oscillator is fed to a wideband amplifier, where the modulation is introduced. Thus, the oscillator is isolated from the modulating signal.

*Frequency or tuning accuracy.* The accuracy of the frequency or tuning dials for a typical shop generator is about 2 or 3%, whereas a lab generator will have from 0.5 to 1% accuracy. However, neither instrument can be used as a frequency standard for servicing CB sets, since the required FCC accuracy is 0.005% or better (preferably 0.0025%). There are laboratory instruments, generally described as communications monitors or frequency-meter/signal-generators, that will provide signals with accuracies of up to 0.00005%. These instruments are designed for commercial communications work (radio and TV broadcast, etc.), are quite expensive, and are thus not usually found in CB service shops!

For practical CB service, the simplest approach is to monitor the signal generator output with a frequency counter. Such counters are discussed in Sec. 2-6. Using this technique, the frequency of the RF signal is determined by the accuracy of the counter, not the generator.

*Frequency range.* Any RF generator should have a range of up to at least 30 MHz. A 50 to 100 MHz frequency range is preferable.

*Frequency drift.* Because a signal generator must provide continuous tuning across a given range, some type of variable-frequency oscillator (VFO) must be used. As a result, the output is subject to drift, instability, modulation (by noise, mechanical shock, or power supply ripple) and other problems associated with VFOs.

Frequency instability does not present too great a problem in practical CB service, provided that you monitor the signal generator output with a frequency counter. Of course, continuous drift can be annoying. For this reason, lab generators incorporate temperature-compensated capacitors to minimize drift. Similarly, the effects of line-voltage variations are offset by regulated power supplies. The better generators also have more elaborate shielding, especially for the output-attenuator circuits, where radio frequency is most likely to leak. The leakage of RF from signal generators is something of a problem during receiver circuit sensitivity tests, or any tests involving low amplitude (microvolt range) RF signals from the generator.

*Band spread.* Shop generators usually have a minimum number of bands for a given frequency range. This makes the tuning-dial or frequency-control adjustments more critical, as well as difficult to see. Lab generators usually have a much greater band spread; that is, they cover a smaller part of the frequency range in each band.

Sec. 2-2  Signal Generators                                                                 71

*Typical RF generators.* Figure 2-1 shows a typical RF signal generator used for CB service (the Heathkit IG-102). The instrument covers 100 kHz to 100 MHz in six bands (100 to 200 MHz on harmonics), with 2% tuning accuracy. Thirty percent internal modulation at 400 Hz is available, as is external modulation.

**Figure 2-1:**  Heathkit IG-102 RF signal generator

An example of a typical lab-type RF generator for CB service is the Hewlett-Packard 606BM. This instrument covers the range from 50 kHz to 65 MHz in six bands with 1% accuracy. Both the modulation and output amplitude are adjustable and are indicated by separate front-panel meters. Internal modulation is at 400 and 1,000 Hz, up to 95%. External modulation is from dc to 20 kHz, again up to 95%.

### 2-2.4  Audio Generators

Audio generators (also known as *audio oscillators*) are useful in troubleshooting the audio/modulation circuits of a CB set. Audio generators may also be used as modulation sources for RF signal generators. For example, if your particular RF generator has only a 400 Hz internal modulation provision, and the CB set literature requires 1,000 Hz (or vice versa), you can modulate the RF generator with an audio generator tuned to 1,000 Hz. In the RF generator shown in Fig. 2-1, you would connect the audio generator output to the EXT. MOD IN connector.

As is the case of RF signal generators, audio generators in their simplest form are essentially audio oscillators. For troubleshooting purposes, the audio output is tunable in frequency over the entire audio range (and beyond) and is variable in amplitude.

Early audio generators produced only sine waves. However, most present-day audio generators also produce square waves at audio frequencies. Some lab audio generators are referred to as *function generators* because they produce various functions: sine waves, square waves, triangular and/or sawtooth waves. Only the sine waves are of any particular value in CB work. However, almost any modern audio generator available today will have some of the other outputs.

The major differences in audio generators are in quality rather than in special features. For example, the better audio generators are less subject to frequency drift and line-voltage variations. The effects of hum or other line noises are minimized by extensive filtering. Accuracy and dial resolution are generally better in lab generators. This makes the tuning-dial adjustments less critical. Lab generators also have a more uniform output over their entire operating range, whereas shop-type generators may vary in amplitude from band to band.

Keep in mind that if you want accuracy from an audio generator, you must monitor the output signal with a meter (for signal amplitude) and a frequency counter (for signal frequency).

### 2-2.5 Pulse Generators

The only practical purpose of a pulse generator in CB service is the test and adjustment of the *noise blanker* circuits. As is discussed in Chapter 3, some CB sets have blanking circuits that detect noise signals or pulses at the receiver input (antenna), and function to desensitize (or cut off) the receiver in the presence of large noise signals. A pulse generator may be used to simulate noise bursts. However, the service literature for every CB set with noise blanking circuits does not always recommend a pulse generator. In many cases, the noise blanking circuits are tested and adjusted with an RF generator, modulated by an AF generator.

## 2-3 OSCILLOSCOPES

There are two uses for oscilloscopes in CB service; signal tracing and modulation measurement.

### 2-3.1 Signal Tracing with an Oscilloscope

The signals in all circuits of a CB set may be traced with an oscilloscope, provided the scope is equipped with the proper probe (probes are

Sec. 2-3 Oscilloscopes 73

discussed further in Sec. 2-5). You can check amplitude, frequency, and waveforms of the signals with a scope. However, many CB service technicians do not use scopes extensively, for the following reasons.

The oscilloscope will measure signal amplitude, but a meter is easier to read. The same applies to signal frequency. The frequency counter is easier to read, and it is far simpler to measure frequency with a meter than with a scope, particularly in the 27 MHz range. The oscilloscope is a superior instrument for monitoring waveforms. However, in CB the signals are mostly sinewave, and waveforms are not critical.

### 2-3.2 Modulation Checks with an Oscilloscope

The main use for an oscilloscope in CB work is to measure percentage of modulation and uniformity or linearity of modulation. The use of an oscilloscope for modulation checks is not new. There are many variations of the basic technique, each of which is discussed below.

### 2-3.3 Direct Measurement of Modulation Envelope with a High-Frequency Oscilloscope

If the vertical channel response of the oscilloscope is capable of handling the transmitter output frequency (at least 30 MHz), the output can be applied through the oscilloscope vertical amplifier. The basic test connections are shown in Fig. 2-2. The procedure is as follows:

1. Connect the oscilloscope to the antenna jack, or the final RF amplifier of the transmitter, as shown in Fig. 2-2. Use one of the three alternatives shown, or the modulation measurement described in the CB set service literature.

2. Key the transmitter (press the push-to-talk switch) and adjust the oscilloscope controls to produce displays as shown. You can either speak into the microphone (for a rough check of modulation), or you can introduce an audio signal (typically at 400 or 1,000 Hz) at the microphone jack input (for a precise check of modulation). Note that Fig. 2-2 provides simulations of typical oscilloscope displays during modulation checks.

3. Measure the vertical dimensions shown as A and B in Fig. 2-2 (the crest amplitude and the trough amplitude). Calculate the percentage of modulation using the equation in Fig. 2-2. For example, if the crest amplitude (A) is 63 (63 screen divisions, 6.3 volts, and so on) and the trough amplitude (B) is 27, the percentage of modulation is:

$$\frac{63-27}{63+27} \times 100 = 40\%$$

Make certain to use the same oscilloscope scale for both the crest (A) and trough (B) measurements.

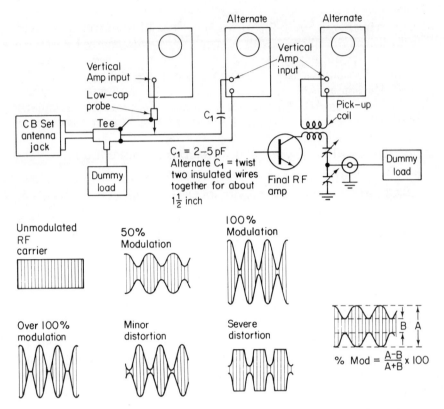

**Figure 2-2:** Direct measurement of modulation envelope with a high frequency (30 MHz or higher) oscilloscope

Keep in mind when making modulation measurements, or any measurement that involves the transmitter, the RF output (antenna connector) must be connected to an antenna or a dummy load. Dummy loads are discussed further in Sec. 2-7.

### 2-3.4 Direct Measurement of Modulation Envelope with a Low-frequency Oscilloscope

If the oscilloscope is not capable of passing 30 MHz, the transmitter output can be applied directly to the vertical deflection plates of the oscilloscope cathode-ray tube. However, there are two drawbacks to this approach. First, the vertical plates may not be readily accessible. Next, the voltage output of the final RF amplifier may not produce sufficient deflection of the oscilloscope trace.

The test connections and modulation patterns are essentially the same as those shown in Fig. 2-2. Likewise, the procedures are the same as those described in Sec. 2-3.3.

## 2-3.5 Trapezoidal Measurement of the Modulation Envelope

The trapezoidal technique has an advantage in that it is easier to measure straight-line dimensions than curving dimensions. Thus, any nonlinearity in modulation may easily be checked with a trapezoid. In the trapezoidal method, the modulated carrier amplitude is plotted as a function of modulating voltage, rather than as a function of time. The basic test connections are shown in Fig. 2-3, and the procedure is as follows:

1. Connect the oscilloscope to the final RF amplifier and modulator. As shown in Fig. 2-3, use either the capacitor connection or the pick-up coil for the RF (oscilloscope vertical input). However, for best results, connect the set outputs directly to the deflection plates of the oscilloscope tube. The oscillo-

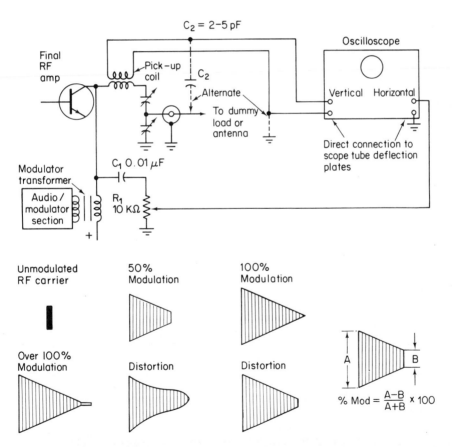

**Figure 2-3:** Trapezoidal measurement of the modulation envelope

scope amplifiers may be nonlinear and will cause the modulation to appear distorted.

2. Key the transmitter and adjust the controls (oscilloscope controls and $R_1$) to produce a display as shown.

3. Measure the vertical dimensions shown as $A$ (crest) and $B$ (trough) on Fig. 2-3, and calculate the percentage of modulation using the equation given. For example, if the crest amplitude ($A$) is 80, and the trough amplitude ($B$) is 40, using the same scale the percentage of modulation is:

$$\frac{80-40}{80+40} \times 100 = 33\%$$

Again, make sure that the transmitter output is connected to an antenna or a dummy load, before transmitting.

### 2-3.6 Down-conversion Measurement of the Modulation Envelope

If the oscilloscope is not capable of passing the 30 MHz carrier signals, and the transmitter output is not sufficient to produce a good indication when connected directly to the oscilloscope tube, it is possible to use a down-converter test setup. One method requires an external RF generator and an IF transformer. The other method uses a receiver capable of monitoring CB channels (such as the receiver of a CB set).

The RF generator method of down-conversion is shown in Fig. 2-4. In this method, the RF generator is tuned to a frequency above or below the

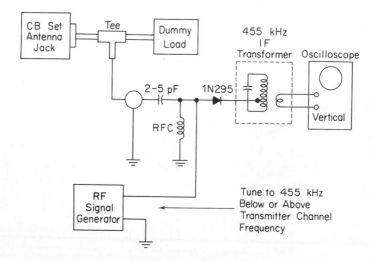

**Figure 2-4:** Down-conversion method of modulation measurement using a 455 kHz IF transformer

Sec. 2-3  Oscilloscopes   77

channel frequency by an amount equal to the IF transformer frequency. For example, if the IF transformer is 455 kHz, tune the RF generator to a frequency 455 kHz above (or below) the channel frequency.

The receiver method of down-conversion is shown in Fig. 2-5. With this method, the receiver is tuned to the channel frequency.

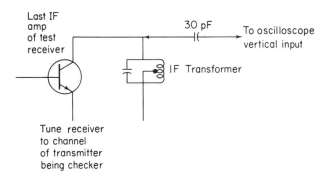

**Figure 2-5:** Down-conversion method of modulation measurement using a CB receiver

In either method, the RF generator or receiver is tuned for a maximum indication on the oscilloscope screen. Once a good pattern is obtained, the rest of the procedure is the same as described in Sec. 2-3.3. The author does not generally recommend the down-conversion methods, except as a temporary measure. There are a number of relatively inexpensive oscilloscopes available that will pass signals up to and beyond the 30 MHz range.

### 2-3.7 Linear Detector Measurement of the Modulation Envelope

If you must use an oscilloscope that will not pass the carrier frequency of 30 MHz, you can use a linear detector. However, the oscilloscope must have a dc input, where the signal is fed directly to the oscilloscope vertical amplifier, not through a capacitor. Most modern-day scopes have both ac (with capacitor) and dc inputs. The basic test connections are shown in Fig. 2-6, and the procedure is as follows:

1. Connect the transmitter output to the oscilloscope through the linear detector circuit as shown in Fig. 2-6. Make certain to include the dummy load.

2. With the transmitter not keyed, adjust the oscilloscope *position* control to place the scope trace on a reference line near the *bottom* of the screen, as shown in Fig. 2-6b.

3. Key the transmitter, but do not apply modulation. Adjust the oscilloscope *gain* control to place the top of the scope trace at the *center* of the screen, as

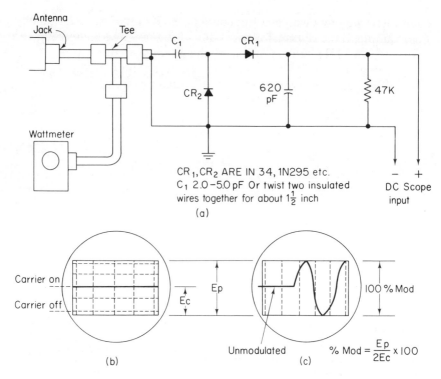

**Figure 2-6:** Pace modulation detector for measurement of the modulation envelope with a low-frequency oscilloscope

shown in Fig. 2-6b. It may be necessary to switch the transmitter off and on several times to adjust the trace properly, since the position and gain controls of most scopes will interact.

4. Measure the distance (in scale divisions) of the shift between the carrier (step 3) and no-carrier (step 2) traces. For example, if the screen has a total of 10 vertical divisions, and the no-carrier trace is at the bottom or zero line, there is a shift of five scale divisions to the centerline.

5. Key the transmitter and apply modulation. Do not touch either the position or gain controls of the oscilloscope.

6. Find the percentage of modulation using the equation shown in Fig. 2-6. For example, assume that the shift between the carrier and no-carrier trace is five divisions, and that the modulation produces a peak-to-peak envelope of eight divisions. The percentage of modulation is:

$$\frac{8}{2 \times 5} \times 100 = 80\%$$

### 2-3.8 Modulation Nomogram

Figure 2-7 is a nomogram that can be used with the direct measurement techniques (Secs. 2-3.3 and 2-3.4) or the trapezoidal technique (Sec. 2-3.5) to find percentage of modulation. To use Fig. 2-7, measure

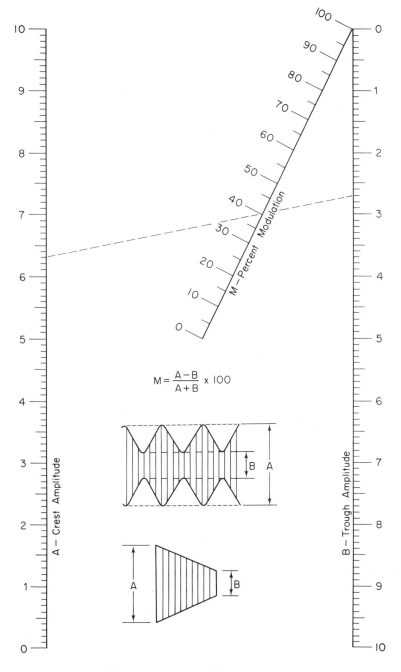

**Figure 2-7:** Pace modulation nomogram

the values of the crest (or maximum) and trough (or minimum) oscilloscope patterns.

The percentage of modulation is found by extending a straight-edge from the measured value of the crest or maximum (given as A on Fig. 2-7) on its scale to the measured value of the trough or minimum (given as B) on its scale. The percentage of modulation is found where the straight-edge crosses the diagonal scale. The crest and trough may be measured in any units (volts, vertical scale divisions, etc.), so long as both crest and trough are measured in the same units. The dashed line in Fig. 2-7 is used to illustrate the percentage of modulation example of Sec. 2-3.3.

## 2-4 METERS

The meters used for CB service are essentially the same as for all other electronic service fields. Most tests can be done with the standard VOM (volt-ohmmeter). The VOM may be either digital (Fig. 2-8) or moving-needle (Fig. 2-9); the choice is yours.

**Figure 2-8:** Heathkit IM-2202 digital multimeter

The digital meter is easier to read, but requires 115 V line power, and is thus best suited to use in the shop. The moving-needle VOM is more difficult to read, but operates on internal batteries, and can thus be used in the shop or in the field. The accuracies of both instruments are about the same. Both meters shown in Figs. 2-8 and 2-9 have more than sufficient accuracy for CB service work.

The meters can be used to measure both voltages and resistances of all CB circuits, as required for the troubleshooting procedures described in Chapters 1 and 4. When used with the appropriate probe (Sec. 2-5), the

Sec. 2-4 Meters

**Figure 2-9:** Heathkit IM-105 moving-needle VOM

meters can be used to trace signals throughout all CB circuits, including receiver (RF and IF), transmitter, and audio/modulator circuits. When used with the correct probe, the meter will indicate the presence of a signal in the circuit and the signal amplitude, but not the signal frequency or waveform.

In addition to accuracy, ranges (both high and low), and resolution or readability, meters are rated in terms of ohms-per-volt; 20,000 ohms-per-volt is typical. A higher ohms-per-volt rating means that the meter draws less current and thus has the least disturbing effect on the circuit under test. A lower ohms-per-volt rating means more circuit loading, which should be avoided in some critical circuits. For example, the AVC or AGC circuits of some CB receivers will not operate properly when loaded with a low ohms-per-volt meter. The same is true of some CB oscillator circuits.

One way to avoid the loading problem is to use an electronic voltmeter that has a high input impedance and thus draws very little current from the circuit under test. The electronic voltmeter can be a VTVM (vacuum-tube voltmeter), EVM (electronic voltmeter), TVM (transistorized voltmeter), or some similar instrument. Most digital meters (Fig. 2-8) are electronic meters and thus draw a minimum of current from the circuit.

One minor problem with some meters is that the frequency range is not sufficient to cover the entire audio range, which is usually considered anything up to about 20 kHz. A typical meter will have a 10 kHz maximum range, without the use of a probe. The problem may be overcome with a probe of the proper type. Also, the range of the audio/modulation circuits in a CB set is 3 kHz maximum. However, some service literature recommends a VTVM, EVM, or TVM that covers the audio range.

## 2-5 PROBES

In practical CB troubleshooting, all meters and oscilloscopes operate with some type of probe. In addition to providing for electrical contact to the circuit being tested, probes serve to modify the voltage being measured to a condition suitable for display on an oscilloscope or readout on a meter.

### 2-5.1 Basic Probe

In its simplest form, the basic probe is a *test prod*. In physical appearance, the probe is a thin metal rod connected to the meter or scope input through an insulated flexible lead. The entire rod, except for the tip, is covered with an insulated handle so that the probe can be connected to any point of the circuit without touching nearby circuit parts. Sometimes, the probe tip is provided with an alligator clip so that it is not necessary to hold the probe at the circuit point.

Such probes work well on CB circuits carrying dc and audio signals. However, if the alternating current is at a high frequency, or if the gain of the meter (such as an electronic meter) or scope amplifier is high, it may be necessary to use a special *low-capacitance probe*. Hand capacitance in a simple probe or test prod can cause hum pickup, particularly if amplifier gain is high. This condition may be offset by shielding in low-capacitance probes. More important, however, is the fact that the input impedance of the meter or scope is connected directly to the circuit being tested when a simple probe is used. Such impedance may disturb circuit conditions (as discussed in Sec. 2-4).

### 2-5.2 Low-capacitance Probes

The basic circuit of a low-capacitance probe is shown in Fig. 2-10. The series resistance $R_1$ and capacitance $C_1$, as well as the parallel or shunt $R_2$, are surrounded by a shielded handle. The values of $R_1$ and $C_1$ are preset at the factory by screwdriver adjustment, and should not be disturbed unless recalibration is required, as discussed in Sec. 2-5.7.

In many low-capacitance probes, the values of $R_1$ and $R_2$ are selected

Sec. 2-5 Probes

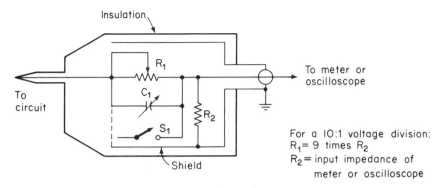

**Figure 2-10:** Typical low-capacitance probe circuit

to form a 10:1 voltage divider between the circuit being tested and the meter or scope input. Thus, the probes serve the dual purpose of capacitance reduction and voltage reduction. You should remember that voltage indications will be one-tenth (or whatever value of attenuation is used) of the actual value when such probes are used. The capacitance value of $C_1$ in combination with the values of $R_1$ and $R_2$ also provide a capacitance reduction, usually in the range of 3:1 to 11:1.

There are probes that combine the features of low-capacitance probes and the basic probe (or test prod). In such probes, a switch (shown as $S_1$ in Fig. 2-10) is used to short both $C_1$ and $R_1$ when a direct input (simple test prod) is required. With $S_1$ open, both $C_1$ and $R_1$ are connected in series with the input, and the probe provides the low-capacitance and voltage-division features.

### 2-5.3 High-voltage Probes

High-voltage probes are rarely, if ever, used in CB service. Most CB sets are solid-state and operate with voltages of less than 15 V. Even the vacuum-tube sets generally use voltages of 300 V or less. Most VOMs will easily handle voltages in this range. However, some meters are supplied with high-voltage probes, either as accessories or built-in. If you should use these probes, the voltage indications will be 10:1, 100:1, or even 1,000:1, depending on the attenuation factor.

### 2-5.4 RF Probes

When the signals to be measured are at radio frequencies and are beyond the frequency capabilities of the meter or scope circuits, an RF probe is required. RF probes convert (rectify) the RF signals to a dc output voltage that is equal to the peak RF voltage. The dc output of the probe is then applied to the meter or scope input and is displayed as a voltage readout in the normal manner. In some RF probes, the dc output is equivalent to peak RF voltage, whereas in other probes, the readout is equal to rms voltage.

The basic circuit of an RF probe is shown in Fig. 2-11. This circuit may be used to provide either peak output or rms output. Capacitor $C_1$ is a high-capacitance dc blocking capacitor used to protect diode $CR_1$.

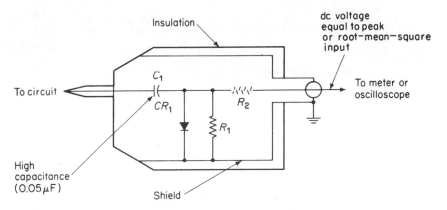

**Figure 2-11:** Typical half-wave RF probe circuit

Usually a germanium diode is used for $CR_1$, which rectifies the RF voltage and produces a dc output across $R_1$. In some probes, $R_1$ is omitted so that the dc voltage is developed directly across the input circuit of the meter or scope. This dc voltage is equal to the peak RF voltage, less whatever forward drop exists across the diode $CR_1$.

When a dc output voltage equal to the root-mean-square of the RF voltage is desired, a series-dropping resistor (shown as $R_2$ in Fig. 2-11) is added to the circuit. Resistor $R_2$ drops the dc output voltage to a level that equals 0.707 of the peak RF value.

The RF probe shown in Fig. 2-11 is a half-wave probe. A full-wave probe provides an output to the meter or scope that is (approximately) equal to the peak-to-peak value of the voltage being measured. This is particularly important when measuring pulses, square waves, and other complex waveforms not generally found in CB work. A full-wave probe circuit is shown in Fig. 2-12. Full-wave probes are usually found with meters rather than scopes. Because most meters are calibrated to read in rms values, the probe is reduced to 0.3535 of the peak-to-peak value. This is done by selecting or adjusting the value or $R_1$. A few electronic voltmeters are provided with peak-to-peak scales, where $R_1$ is omitted. These can be used for CB, but are better suited to TV and laboratory applications.

Sec. 2-5   Probes

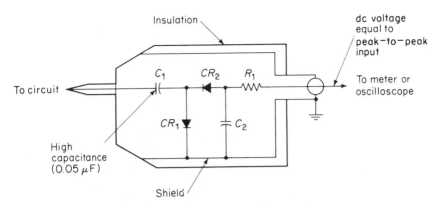

Figure 2-12:  Typical full-wave RF probe circuit

### 2-5.5   Demodulator Probes

The circuit of a demodulator probe is essentially the same as that of the RF probe, but the circuit values and the basic functions are somewhat different. Both the half-wave and full-wave probes are used to convert the high-frequency signals (usually an RF carrier) to a dc voltage that can be measured on a meter or scope. When the high-frequency signals contain modulation (such as a modulated RF carrier), a demodulator probe is more effective for signal tracing.

The basic circuit of the demodulator probe is shown in Fig. 2-13. Here, capacitor $C_1$ is a low-capacitance dc blocking capacitor. (In the RF probe, a high-capacitance value is required to ensure that the diode operates at the peak of the RF signal. This is not required for a demodulator probe.) Germanium diode $CR_1$ demodulates (or detects) the AM

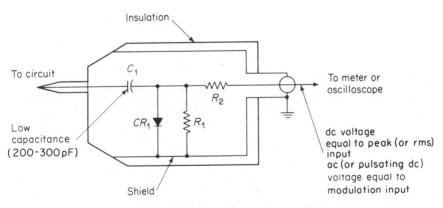

Figure 2-13:  Typical demodulator probe circuit

signal and produces voltages across load resistor $R_1$. ($C_1$ and $R_1$ also act as a filter.)

The demodulator probe produces ac and dc outputs. The RF carrier frequency is converted to a dc voltage equal to the peak of the RF carrier. The low-frequency modulating voltage appears as ac at the probe output.

In troubleshooting, the meter or scope is set to measure direct current, and the RF carrier is measured. Then the meter or scope is set to measure alternating current, and the modulating voltage is measured. Resistor $R_2$ is used primarily for isolation between the circuit being tested and the meter or scope input. Resistor $R_2$ can also serve as a calibrating resistance, so that the output will be of equal value (root-mean-square, peak, and so on). However, in general, demodulator probes are used primarily for *signal tracing* (as part of troubleshooting), and their output is not calibrated to any particular value.

### 2-5.6 Transistorized Signal-tracing Probes

It is possible to increase the sensitivity of a probe with a transistor amplifier. Such an arrangement is particularly useful with a basic VOM for measuring small signal voltages during troubleshooting. An amplifier is usually not required for an electronic meter or scope because such instruments contain built-in amplifiers.

A transistor probe and amplifier circuit is shown in Fig. 2-14. This circuit will increase the sensitivity of the probe by at least 10:1 and should provide good response up to about 500 MHz. The circuit is not normally calibrated to provide a specific voltage indication; rather, it is used to increase the sensitivity of the probe for signal-tracing purposes.

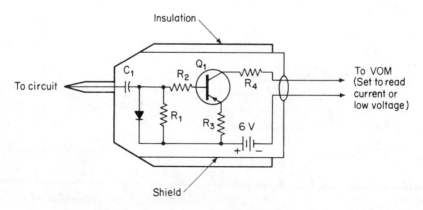

**Figure 2-14:** Typical transistorized signal-tracing probe circuit

## 2-5.7 Probe Compensation and Calibration

Probes must be calibrated to provide a proper output to the meter or scope with which they are to be used. Probe compensation and calibration are done at the factory and require proper test equipment. The following paragraphs describe the *general* procedures for compensating and calibrating probes. *Never* attempt to adjust a probe unless you follow the instruction manual and have the proper test equipment. An improperly adjusted probe will produce erroneous readings and may cause undesired circuit loading.

*Probe compensation.* The capacitors that compensate for excessive attenuation of high-frequency signal components (through the probe's resistance dividers) affect the entire frequency range from some midband point upward. Capacitor $C_1$ in Fig. 2-10 is an example of such a compensating capacitor.

Compensating capacitors must be adjusted so that the higher-frequency components are attenuated by the same amount as low frequency and direct current. It is possible to check the adjustment of the probe-compensating capacitors using a square-wave signal source. This is done by applying the square-wave signal directly to a scope input and then applying the same signals through the probe and noting any change in pattern. In a properly compensated probe, there should be no change (except for a possible reduction of the amplitude).

Figure 2-15 shows typical square-wave displays with the probe properly compensated, undercompensated (high frequencies underemphasized), and overcompensated (high frequencies overem-

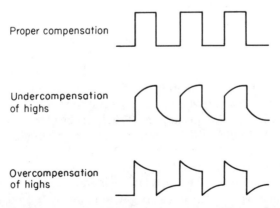

**Figure 2-15:** Typical squarewave displays showing frequency compensation of probes

phasized). Proper compensation of probes is often neglected, especially when probes are used interchangeably with meters or scopes having different input characteristics. It is recommended that any probe be checked with square-wave signals before it is used in troubleshooting.

Another problem related to probe compensation is that the input capacitance of the meter or scope may change with age. Also, in the case of vacuum-tube meters and scopes, the input capacitance may change when tubes are changed. Either way, the compensated dividers may be improperly adjusted. Readjustment of the probe will not correct for the change needed by the input circuits of the meter or scope.

*Probe calibration.* The main purpose of probe calibration is to provide a specific output for a given input. For example, the value of $R_1$ in Fig. 2-12 is adjusted (or selected) to provide a specific amount of voltage to the meter or scope. During calibration, a voltage of known value and accuracy is applied to the input. The output is monitored, and $R_1$ is adjusted to produce a given value (0.707 of RF peak value, etc.).

### 2-5.8 Probe Troubleshooting Techniques

Although a probe is a simple instrument and does not require specific operating procedures, several points should be considered in order to use a probe effectively in troubleshooting.

*Circuit loading.* When a probe is used, the probe's impedance (rather than the meter's or the scope's impedance) determines the amount of circuit loading. As discussed, connecting a meter or scope to a circuit may alter the signal at the point of connection. To prevent this, the impedance of the measuring device must be large in relation to that of the circuit being tested. Thus, a high-impedance probe will offer less circuit loading, even though the meter or scope may have a lower impedance.

*Measurement error.* The ratio of the two impedances (of the probe and the circuit being tested) represents the amount of probable error. For example, a ratio of 100:1 (perhaps a 100 MΩ probe to measure the voltage across a 1 MΩ circuit) will account for an error of about 1%. A ratio of 10:1 will produce an error of about 9%.

*Effects of frequency.* The input impedance of a probe is not the same at all frequencies; it continues to get smaller at higher frequencies. (Capacitive reactance and impedance decrease with an increase in frequency.) All probes have some input capacitance. Even an increase at audio frequencies may produce a significant change in impedance.

Sec. 2-5   Probes                                                                89

*Shielding capacitance.* When using a shielded cable with a probe to minimize pickup of stray signals and hum, the additional capacitance of the cable should be considered. The capacitance effects of a shielded cable can be minimized by terminating the cable at one end in its characteristic impedance. Unfortunately, this is not always possible with the input circuits of most meters and scopes.

*Relationship of loading to attenuation factor.* The reduction of loading (either resistive or capacitive) due to use of probes may not be the same as the attenuation factor of the probe. (Capacitive loading is almost never reduced by the same amount as the attenuation factor because of the additional capacitance of the probe cable.) For example, a typical 5:1 attenuator probe may be able to reduce capacitive loading by about 2:1. A 50:1 attenuator probe may reduce capacitive loading by about 10:1. Beyond this point, little improvement can be expected because of stray capacitance at the probe tip.

*Checking effects of the probe.* When troubleshooting, it is possible to check the effect of a probe on a circuit by making the following simple test: attach and detach another connection of similar kind (such as another probe) and observe any difference in meter reading or scope display. If there is little or no change when the additional probe is touched to the circuit, it is safe to assume that the probe has little effect on the circuit.

*Probe length and connections.* Long probes should be restricted to the measurement of relatively slow-changing signals (direct current and low-frequency ac). The same is true for long ground leads. The ground leads should be connected where no hum or high-frequency signal components exist in the ground path between that point and the signal-pickoff point.

*Measuring high voltages.* Avoid applying more than the rated voltage to a probe. Fortunately, most commercial probes will handle the highest voltages found in CB, even vacuum-tube CB sets.

### 2-5.9   An RF Probe for CB Service

Figure 2-16 shows the schematic diagram (with circuit values) of a probe suitable for CB service. The probe is designed specifically for use with a VOM or electronic voltmeter, and will convert AF and RF signals to direct current. The RF signals can be well above the 30 MHz range of a CB set. (The probe will work satisfactorily up to 250 MHz and above.) Note that the 47 k$\Omega$ resistor is not used with a VOM like that shown in

Fig. 2-9. Also, the probe is essentially a signal-tracing device and is not designed to provide accurate readings. The meter used with the probe must be set to read direct current, because the probe output is dc. However, if the RF input signal is modulated, the probe output may be pulsating direct current.

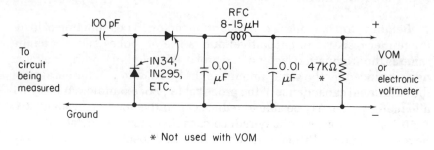

* Not used with VOM

**Figure 2-16:** Pace RF probe for CB service

## 2-6 FREQUENCY METERS AND COUNTERS

There are two basic types of frequency measuring devices for CB service: the heterodyne or zero-beat frequency meter and the digital electronic counter.

### 2-6.1 Heterodyne or Zero-beat Frequency Meter

In the early days of radio communications, the heterodyne meter was the only practical device for frequency measurement of transmitter signals. Figure 2-17 shows the block diagram of a basic heterodyne

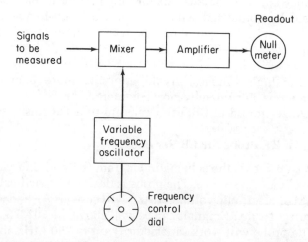

**Figure 2-17:** Basic heterodyne or zero-beat frequency meter

Sec. 2-6 Frequency Meters and Counters

frequency meter. The signals to be measured are applied to a mixer, along with signals of known frequency (usually from a variable-frequency oscillator in the meter). The meter oscillator is adjusted until there is a null or "zero-beat" on the output device, indicating that the oscillator is at the same frequency as the signals to be measured. This frequency is read from the oscillator frequency control dial. Precision frequency meters often include charts or graphs to help interpret frequency dial readings, so that exact frequencies can be pinpointed.

As an alternate system, the meter produces fixed frequency signals that are applied to the mixer, together with the signals of unknown frequency. As an example, one such frequency meter provides 23 crystal-controlled signals, one signal for each of the 23 CB channels. Both the CB transmitter and frequency meter are set to the same channel, and any deviation is read out on the frequency meter indicator.

### 2-6.2 Electronic Digital Counter

The electronic counter has become far more popular for CB work than the heterodyne frequency meter. One reason is that the counter is generally easier to operate and has much greater resolution or readability. Using the counter, you need only connect the test leads to the circuit or test point, select a time base and attenuator/multiplier range, and read the signal frequency on a convenient digital readout.

The electronic counter shown in Fig. 2-18 is typical for CB service (the Heathkit IM-4100). As a frequency *counter*, the device will read up to 30 MHz. In the *period* mode, the instruments measure intervals up to 99.999 seconds. In *totalize* operation, the instrument will add up (totalize) events up to a count of 99,999. Neither the period or totalize

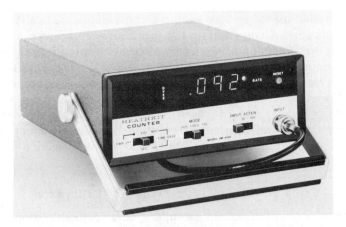

**Figure 2-18:** Heathkit IM-4100 frequency counter

modes are particularly important in CB service, but they do expand the instrument's capability. An example of a typical lab-type frequency counter for CB service is the Hewlett-Packard 4245L.

*Counter accuracy.* The accuracy of a frequency counter is set by the stability of the time base, rather than the readout. The readout is typically accurate to within ±1 count. The time base of the Fig. 2-18 counter is 10 MHz, and is stable to within ±10 ppm (parts per million), or 100 Hz. The time base of a precision lab counter could be in the order of 4 MHz and is stable to within ±1 ppm, or 4 Hz.

*Counter resolution.* The resolution of an electronic counter is set by the number of digits in the readout. For example, assume that you measure a 27 MHz signal with a five-digit counter. The count could be 26.999 or 27.001, or within 1,000 Hz of 27 MHz. Since the FCC requires that the frequency be held within 0.005 % (or about 1350 Hz in the case of a 27 MHz signal), a digital counter for CB work must have a *minimum* of five digits in the readout.

*Combining accuracy and resolution.* To find out if a counter is adequate for CB service, add the time base stability (in terms of frequency) to the resolution at 27 MHz. For example, if the accuracy is 100 Hz, and the count can be resolved to 1,000 Hz (at the measurement frequency), the maximum possible inaccuracy is 1,000+100 Hz, or 1,100 Hz. This is within the approximate, 1,350 Hz required.

### 2-6.3 Calibration Check of Frequency Meters and Counters

The accuracy of frequency measuring devices (both meters and counters) used for CB service should be checked periodically, at least every 6 months. Always follow the procedures recommended in the frequency meter or counter service instructions. Generally, you can send the instrument to a calibration lab, or to the factory, or you can maintain your own frequency standard. (This latter is generally not practical for most CB service shops.)

No matter what standard is used, keep in mind that the standard must be more accurate, and have better resolution, than the frequency measuring device, just as the meter or counter must be more accurate than the CB set.

### 2-6.4 Using WWV Signals for Frequency Calibration

In the absence of a frequency standard, or factory calibration, you can use the frequency information broadcast by U. S. Government station WWV. These WWV signals are broadcast on 2.5, 5, 10, 15, 20, and 25

Sec. 2-6  Frequency Meters and Counters                                     93

MHz continuously night and day, except for silent periods of approximately 4 minutes beginning 45 minutes after each hour. Broadcast frequencies are held accurate to within 5 parts in $10^{11}$. This is far more accurate than that required for CB work. The hourly broadcast schedules of WWV are shown in Fig. 2-19. However, these schedules are subject to change. For full data on WWV broadcasts, refer to NBS (National Bureau of Standards) Standard Frequency and Time Services (Miscellaneous Publication 236), available from Superintendent of Documents, U. S. Government Printing Office, Washington, D.C. 20402.

It is the CW (continuous wave) signals broadcast by WWV that provide the most accurate means of calibrating (or checking) frequency meters and counters. It is not practical to use the signal directly, except on some special frequency meters, but the test connections for check are not complex.

Figure 2-20 shows the basic test connections for checking the accuracy of a frequency counter using WWV. Note that a receiver and signal generator are required. The accuracy of the signal generator and receiver are not critical, but both instruments must be capable of covering the desired frequency range. The procedure is as follows:

1. Allow the signal generator, receiver, and counter being tested to warm up for at least 15 minutes.

2. Reduce the signal generator output amplitude to zero. Turn off the signal generator output, if this is possible without turning off the entire signal generator.

3. Tune the receiver to the desired WWV frequency. You may use all WWV frequencies, but the 25 MHz frequency is most useful, since it is near the CB frequency of 27 MHz.

4. Operate the receiver controls until you can hear the WWV signal in the receiver loudspeaker.

5. If the receiver is of the communications type, it will have a BFO (beat-frequency oscillator) and output signal strength or S-meter. Turn on the BFO, if necessary, to locate and identify the WWV signal. Then tune the receiver for maximum signal on the S-meter. The receiver is now exactly on 25 MHz, or whatever WWV frequency is selected.

6. Turn on the signal generator, and tune the generator until it is at "zero-beat" against the WWV signal. As the signal generator is adjusted close to that of the WWV signal (so that the difference in frequency is within the audio range) a tone, whistle, or "beat note" will be heard on the receiver. When the signal generator is adjusted to exactly the WWV frequency (25 MHz in this case), there is no "difference" signal, and the tone can no longer be heard. In effect the tone drops to zero and the two signals (generator and WWV) are at "zero beat."

7. Read the counter. The indication should be equal to the WWV frequency. For example, with a five-digit counter at 25 MHz, the reading should be 24.999 to 25.001.

8. Repeat the procedure at other WWV broadcast frequencies.

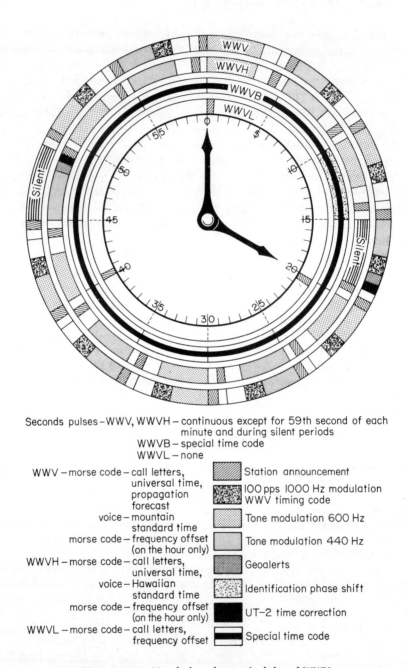

Seconds pulses – WWV, WWVH – continuous except for 59th second of each minute and during silent periods
WWVB – special time code
WWVL – none

WWV – morse code – call letters, universal time, propagation forecast
voice – mountain standard time
morse code – frequency offset (on the hour only)

WWVH – morse code – call letters, universal time,
voice – Hawaiian standard time
morse code – frequency offset (on the hour only)

WWVL – morse code – call letters, frequency offset

- Station announcement
- 100 pps 1000 Hz modulation WWV timing code
- Tone modulation 600 Hz
- Tone modulation 440 Hz
- Geoalerts
- Identification phase shift
- UT-2 time correction
- Special time code

**Figure 2-19:** Hourly broadcast schedules of WWV

Sec. 2-7 Dummy Load

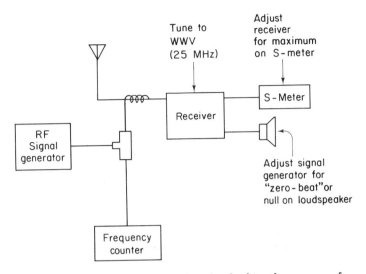

**Figure 2-20:** Basic test connections for checking the accuracy of a frequency counter using WWV

## 2-7 DUMMY LOAD

Never adjust the transmitter of a CB when the set is connected to a radiating antenna! You are almost certain to cause interference. Equally important, never adjust the transmitter without an antenna or load connected to the output. This will probably cause interference and will almost certainly cause damage to the set. When a transmitter is connected to an antenna or load, power is transferred from the final RF stage (vacuum-tube or transistor) to the antenna or load. Without an antenna or load, the final RF stage must dissipate the full power and will probably be damaged.

These problems may be overcome by means of a nonradiating load, commonly called a dummy load. There are a number of commercial dummy loads for CB service. The RF wattmeters described in Sec. 2-8 and the special test sets covered in Sec. 2-14 contain dummy loads. It is also possible to make up dummy loads suitable for most CB service procedures. There are two generally acceptable dummy loads: the fixed resistance and the lamp. Keep in mind that these loads are for routine service work; they are not a substitute for an RF wattmeter or test set.

### 2-7.1 Fixed Resistor Dummy Load

The simplest dummy load is a fixed resistor capable of dissipating the full power output of the transmitter. The resistor can be connected to the set's antenna connector by means of a plug, as shown in Fig. 2-21.

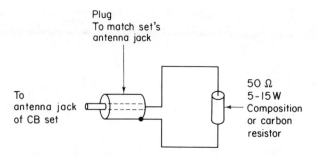

**Figure 2-21:** Fixed resistor dummy load

Most CB sets operate with a 50 Ω antenna and lead-in, and thus require a 50 Ω resistor. The nearest standard resistor is 51 Ω. This 1 Ω difference is not critical. However, it is essential that the resistor be noninductive (composition or carbon), never wire-wound. Wire-wound resistors have some inductance, which changes with frequency; thus, the load (impedance) presented by the resistor changes with frequency. The fixed resistor should be capable of dissipating a minimum of 5 W for AM sets, and 15 W for SSB sets.

*RF power output measurement with dummy load resistor.* It is possible to get an approximate measurement of RF power output from a CB transmitter with a resistor dummy load and a suitable meter. Again, these procedures are not to be considered a substitute for power measurement with an accurate RF wattmeter (Sec. 2-8).

The procedure is simple. Measure the voltage across the dummy load resistor and find the power with the equation:

$$\text{power} = \frac{(\text{voltage})}{50}$$

For example, if the voltage measured is 14 V, the power output is:

$$\frac{(14)^2}{50} = 3.92\,\text{W}$$

Certain precautions must be observed. First, the meter must be capable of producing accurate voltage indications at 27 MHz. This will probably require a meter with probe. An AM set should be checked with an rms voltmeter and with no modulation applied. An SSB set must be checked with a peak-reading voltmeter and with modulation applied (since SSB produces no output without modulation). This usually involves connecting an audio generator to the microphone input of an SSB set.

Sec. 2-8    RF Wattmeter    97

Always follow the service literature recommendations for all RF power output measurements (frequency, channels, operating voltages, modulation, etc.). However, as a guideline, an AM set should produce no more than 4 W output with or without modulation. SSB sets should not produce more than 12 W output with full modulation (speaking loudly into the microphone).

### 2-7.2  Lamp Dummy Load

Lamps have been the traditional dummy loads for communications service work. The No. 47 lamp (often found as a pilot lamp in many electronic instruments) provides the approximate impedance and power dissipation required as a CB dummy load. The connections are shown in Fig. 2-22.

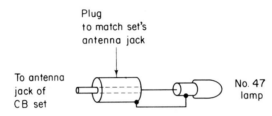

**Figure 2-22:**  Lamp (NO. 47) dummy load

You cannot get an accurate measurement of RF power output when a lamp is used as the dummy load. However, the lamp will provide an indication of relative power and will show the presence of modulation. The intensity of the light produced by the lamp will vary with modulation (more modulation, brighter glow). Thus, you can tell at a glance if the transmitter is producing an RF carrier (steady glow), and if modulation is present (varying glow).

## 2-8  RF WATTMETER

A number of commercial RF wattmeters are available for CB service work. Also, the special test sets described in Sec. 2-14 usually include an RF wattmeter. The basic RF wattmeter consists of a dummy load (fixed resistor) and a meter that measures voltage across the load but reads out in watts (rather than volts). You simply connect the RF wattmeter to the antenna connector of the set (transmitter output), key the transmitter, and read the power output on the wattmeter scale.

Although operation is simple, you must remember that SSB sets require a peak-reading wattmeter to indicate PEP (peak envelope power), whereas an AM set uses an rms-reading wattmeter. Most RF wattmeters are rms-reading, unless specifically designed for SSB.

## 2-9 VACUUM TUBE, TRANSISTOR, CAPACITOR, AND CRYSTAL TESTERS

The testers used in CB service are essentially the same as those used in other electronic fields. That is, the same testers may be used, in the same way and for the same purpose. Thus, we will not discuss such testers in any great detail.

### 2-9.1 Vacuum-tube Testers

Tube testers are not popular test instruments in most CB shops, since tubes are being replaced by transistors, and the need for such testers is being eliminated. As discussed in Chapter 1, the only test of a vacuum tube is its performance in-circuit. This is particularly true of tubes used in the high-frequency circuits (oscillator, RF amplifiers, etc.). If you remove the tube for test and find it defective on the tester, you must substitute a new tube in the set. You may use substitution as the first step. The choice of a tube tester is up to you. If you do decide to use a tester, the transconductance type is generally the best.

### 2-9.2 Transistor Testers

Transistors may be tested in or out of circuit using commercial transistor testers. These testers are the solid-state equivalent of vacuum-tube testers (although they do not operate on the same principle).

The use of transistor testers in CB troubleshooting is generally a matter of individual preference. At best, such testers show the gain and leakage of transistors at direct current or low frequencies under one set of conditions (fixed voltage, current, and so forth).

*Typical commercial transistor tester.* It is impractical here to discuss every type of transistor test circuit available. One very effective method for both in-circuit and out-of-circuit tests is shown in Fig. 2-23. In this method, a 60 Hz square-wave pulse is applied simultaneously to the base-emitter and the collector-emitter junctions. The current flow in each of the two junctions is measured and compared. The difference in current flow (collector-emitter divided by base-emitter) is the transistor gain.

It is not necessary to remove the transistor from the circuit to make such a test, but the set must be turned off. In fact, this test method will also show up defects in the transistor circuit. For example, in many transistor circuits, the overall gain is set by the circuit-resistance values rather than by transistor gain. Assume that a particular circuit has

Sec. 2-9  Vacuum Tube, Transistor, Capacitor, and Crystal Testers   99

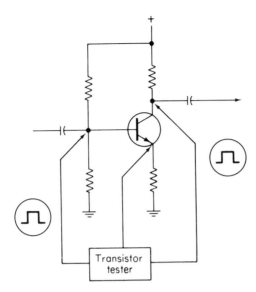

**Figure 2-23:** In-circuit transistor test method

resistance values that would normally show a gain of 10, and that the in-circuit square-wave test (using a tester) shows no gain or very low gain. This could be the result of a bad transistor or circuit problems, or both. If the transistor is then tested out of circuit under identical conditions and a gain is shown, the problem is most likely one of an undesired change in circuit-resistance values.

### 2-9.3  Capacitor Tester

A capacitor tester will establish that the capacitor value is correct. If the tester shows the value to be good, it is reasonable to assume that the capacitor is not open, shorted, or leaking. From another standpoint, if a capacitor shows no opens, shorts, or leakage, it is also reasonable to assume that the capacitor is good. Thus a simple test that shows the possibility of shorts, opens, or leakage is usually sufficient. These procedures may be accomplished with a VOM, and are described in Chapter 4.

### 2-9.4  Crystal Tester

Crystal testers or checkers used in CB service are generally of the "crystal activity" type. The tester consists of an oscillator circuit and a meter connected to measure current in the circuit (usually collector current). The meter indicates the presence of oscillation in the crystal in the tester circuit. In some testers, the meter is adjusted to a "set" or "calibrate" position, the crystal is removed, and the meter needle then moves to a "good" or "bad" position. Some of the special test sets described in Sec. 2-14 include a crystal activity test function.

### 2-9.5 Diode Testers

There are two basic types of diode testers: the laboratory type (generally not used in CB service) and the simple, battery-operated, portable type. These simple testers check diode operation out-of-circuit on a good-bad basis. Usually, the diodes are checked on the basis of forward and reverse resistance, or the resistance ratio. These tests can be performed with a VOM, and are described in Chapter 4.

## 2-10 FIELD STRENGTH METER

There are two basic types of field strength meters: the simple relative field strength (RFS) meter and the precision laboratory or broadcast-type instrument. Most CB service procedures can be carried out with simple RFS instruments (unless you must make precision measurements of broadcast antenna radiation patterns).

The purpose of a field strength meter is to measure the strength of signals radiated by an antenna. This simultaneously tests the transmitter output and antenna lead-in. In its simplest form, a field strength meter consists of an antenna (a short piece of wire or rod), a potentiometer, diodes, and a microammeter, as shown in Fig. 2-24. More elaborate RFS meters include a tuned circuit and possibly a transistor amplifier. In use, the meter is placed near the antenna or at some location accessible to the

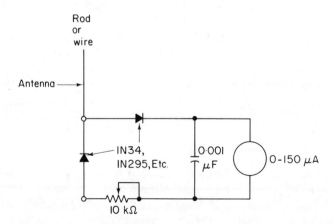

**Figure 2-24:** Basic relative field strength (RFS) meter circuit

Sec. 2-11 Standing Wave Ratio (SWR) Measurement                   101

CB set (where you can see the meter), the transmitter is keyed, and the *relative* field strength is indicated on the meter. Some of the special test sets described in Sec. 2-14 include an RFS test function.

## 2-11 STANDING WAVE RATIO (SWR) MEASUREMENT

The standing wave ratio (SWR) of an antenna is actually a measure of match or mismatch for the antenna, transmission line (lead-in), and the CB set. When the impedances of the antenna, line, and set are perfectly matched, all of the energy or signal will be transferred to or from the antenna, and there will be no loss. If there is a mismatch (as is the case in any practical application), some of the energy or signal will be reflected back into the line. This energy will cancel part of the desired signal.

If the voltage (or current) is measured along the line, there will be voltage or current maximums (where the reflected signals are in-phase with the outgoing signals) and voltage or current minimums (where the reflected signal is out-of-phase, partially canceling the outgoing signal). The maximums and minimums are called *standing waves*. The ratio of the maximum to the minimum is the standing wave ratio, or SWR. The ratio may be related to either voltage or current. Since voltage is easier to measure, it is usually used, resulting in the common term *voltage standing wave ratio*, or VSWR. The theoretical calculations for VSWR are shown in Fig. 2-25.

An SWR of 1-to-1, expressed as 1:1, means that there are no maximums or minimums (the voltage is constant at any point along the line) and that there is a perfect match for set, line, and antenna. As a practical matter, if this 1:1 ratio should occur on one CB channel, or one frequency, it will not occur at any other frequency, since impedance

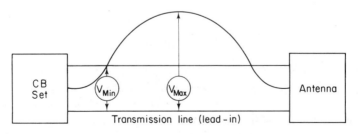

Voltage standing wave ratio VSWR = $\dfrac{V_{Max}}{V_{Min}}$

**Figure 2-25:** Calculations for voltage standing wave ratio (VSWR)

changes with frequency. It is not likely that all three elements (set, antenna, line) will change impedance by the exact same amount on all channels. Therefore, when checking SWR, always check on all channels, where practical. As an alternative, check SWR at the high, low, and middle channels.

In the case of microwave signals, a meter is physically moved along the line to measure maximum and minimum voltages. This is not practical at CB frequencies, due to the physical length of the waves. In CB work, it is more practical to measure forward or outgoing voltage and reflected voltage and then calculate the *reflection coefficient* (reflected voltage/outgoing voltage). The relationship of reflection coefficient to SWR is as follows:

$$\text{Reflection coefficient} = \frac{\text{reflected voltage}}{\text{forward voltage}}$$

For example, using a 10 V forward and the 2 V reflected, the reflection coefficient is 0.2.

Reflection coefficient is converted to SWR by dividing (1+reflection coefficient) by (1−reflection coefficient). For example, using the 0.2 reflection coefficient, the SWR is:

$$\frac{1+0.2}{1-0.2} = \frac{1.2}{0.8} = 1.5 \text{ SWR}$$

This may be expressed as 1 to 1.5, 1.5 to 1, 1.5:1, 1:1.5, or simply as 1.5, depending the meter scale. In practical terms, an SWR of 1.5 is poor, since it means that at least 20% of the power is being reflected.

SWR can be converted to reflection coefficient by dividing (SWR−1) by (SWR+1). For example, using the 1.5 SWR, the reflection coefficient is:

$$\frac{1.5-1}{1.5+1} = \frac{0.5}{2.5} = 0.2 \text{ reflection coefficient}$$

In the commercial SWR meters used for CB work, it is not necessary to calculate either reflection coefficient or SWR. This is done automatically by the SWR meter. That is, the meter is actually reading reflection coefficient, but the *scale reads out* in SWR. (If you have a reflection coefficient of 0.2, the SWR reading will be 1.5.)

There are a number of SWR meters used in CB service. Some CB sets even have a built-in SWR measurement circuit. The SWR function is

Sec. 2-11   Standing Wave Ratio (SWR) Measurement                               **103**

often combined with other measurement functions (field strength, power output, etc.). Practically all of the special test sets described in Sec. 2-14 include an SWR measurement feature, since it is so important to proper operation of a CB set.

Basic SWR meter circuits are quite simple, and it is possible to fabricate them in the shop. However, it is not practical in most cases to do so. The basic circuit requires that a *directional coupler* and pickup wires be inserted in the transmission line. Even under good conditions, a mismatch and some power loss may result. A poorly designed and fabricated pickup may result in considerable power loss, as well as inaccurate readings. Thus, it is more practical to use commercial SWR meters.

The basic SWR meter circuit is shown in Fig. 2-26. Operation of the circuit is as follows. As shown, there are two pickup wires, both parallel to the center conductor of the transmission line. Any RF voltage on

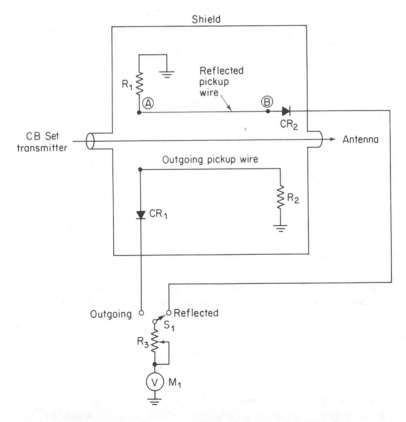

**Figure 2-26:** Basic SWR meter circuit (directional coupler)

either of the parallel pickups is rectified and applied to the meter through switch $S_1$. Each pickup wire is terminated in the impedance of the transmission line by corresponding resistors $R_1$ and $R_2$ (typically 50 to 52 $\Omega$).

The outgoing voltage (CB transmitter to antenna) is absorbed by $R_1$. Thus, there is no outgoing voltage on the reflected pickup wire beyond point A. However, the outgoing voltage remains on the transmission line at the outgoing pickup wire. This voltage is rectified by $CR_1$, and appears as a reading on the meter, when $S_1$ is in the outgoing voltage position.

The opposite condition occurs for the reflected voltage (antenna to transmitter). There is no reflected voltage on the outgoing pickup wire beyond point B, because the reflected voltage is absorbed by $R_2$. The reflected voltage does appear on the reflected pickup wire beyond this point and is rectified by $CR_2$. The reflected voltage appears on meter $M_1$ when $S_1$ is in the reflected voltage position.

In use, switch $S_1$ is set to read outgoing voltage, and resistor $R_3$ is adjusted until the meter needle is aligned with some "set" or "calibrate" line (near the right-hand end of the meter scale). Switch $S_1$ is then set to read reflected voltage, and the meter needle moves to the SWR indication.

As a practical matter, SWR meters often do not read beyond 1:3. This is because an indication above 1:3 indicates a poor match. Make certain you understand the scale used on the SWR meter. For example, a typical SWR meter will be rated at 1:3, meaning that it will read SWR from 1:1 to 1:3. However, the scale indications are 1, 1.5, 2, and 3. These scale indications mean 1:1, 1:1.5, 1:2, and 1:3, respectively. The scale indications between 1 and 1.5 are the most useful, since a good antenna system will show a typical 1.1 or 1.2. Anything between 1.2 and 1.5 is on the borderline.

## 2-12 DIP METERS

The dip meter, or grid-dip meter, has long been a tool in radio communications service work, particularly in amateur radio. The dip meter has many uses, but its most useful function in CB work is in presetting "cold" transmitter and receiver resonant circuits (no power applied to the CB set). This makes it possible to adjust the resonant circuits of a badly-tuned set, or a set where new coils and transformers must be installed as a replacement.

As an example, it is possible that the replacement coil or transformer will be tuned to an undesired frequency when shipped from the factory. Using a dip meter, it is possible to install the coil, tune it to the correct

Sec. 2-12   Dip Meters                                                                 **105**

frequency, and then apply power to the set and adjust the circuit for "peak" as described in the service literature. (Most service literature assumes that the circuits are not badly tuned and only require "peaking.")

There are many types of dip meters and circuits. A typical dip meter is a hand-held, battery-operated device as shown in Fig. 2-27. The circuit is essentially an RF oscillator with external coil, a tuning dial, and a meter. When the coil is held near the circuit to be tested, and the oscillator is tuned to the resonant frequency of the test circuit, part of the RF energy is absorbed by the test circuit, and the meter "dips." The procedure can be reversed, where the dip meter is set to a desired frequency, and the test circuit is tuned to produce a "dip" on the meter.

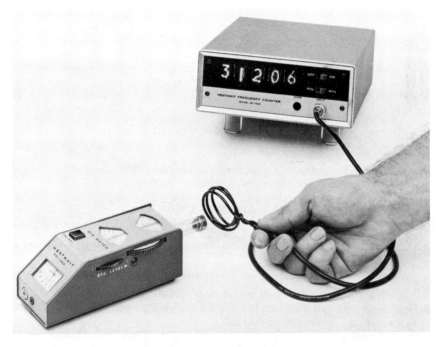

**Figure 2-27:** Heathkit HD-1250 dip meter used with Heathkit IB-1100 frequency counter

We will not go into the many uses of the dip meter here. Such subjects are covered thoroughly in the author's *Handbook of Practical Electronic Tests and Measurements*, Prentice-Hall, Inc., Englewood Cliffs, New Jersey. Instead, we shall describe how a *dip adapter* may be used to preset resonant circuits or to check the frequency of resonant circuits.

### 2-12.1 The Basic Dip Adapter

A basic dip adapter circuit is shown in Fig. 2-28. Such a circuit may be fabricated in the shop with little difficulty. Resistor $R_1$ should match the impedance of the signal generator (typically 50 Ω). Both diode $CR_1$ and the microammeter should match the output of the signal generator. The pick-up coil $L_1$ consists of a few turns of insulated wire. The accuracy of the dip adapter circuit depends on the counter accuracy, or on the signal generator accuracy if the counter is omitted.

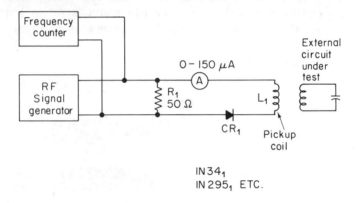

**Figure 2-28:** Basic dip adapter circuit

### 2-12.2 Setting Resonant Frequency with a Dip Adapter

The frequency of a resonant circuit may be set using a dip adapter. The following procedure is applicable to both series and parallel resonant circuits.

1. Couple the dip adapter to the resonant circuit using pickup coil $L_1$ of Fig. 2-28. Usually, the best coupling has a few turns of $L_1$ passed over the coil of the resonant circuit. Make certain that the CB set is off.
2. Set the signal generator to the desired resonant frequency, as indicated by the frequency counter. Adjust the signal generator output amplitude control for a convenient reading on the adapter meter.
3. Tune the resonant circuit for a maximum dip on the adapter meter. The resonant circuits of CB sets may be tuned by means of adjustable slugs in the coil and/or adjustable capacitors.
4. Most resonant circuits are designed so as not to tune across both the fundamental resonant frequency and any harmonics. However, it is possible that the circuit will tune to a harmonic and produce a dip. To check this condition, tune the resonant circuit for maximum dip, and set the signal generator to the first harmonic (twice the desired resonant frequency) and to the first subhar-

monic (one-half the resonant frequency). Note the amount of dip at both harmonics. The harmonics should produce substantially smaller dips than the fundamental resonant frequency.

5. For maximum accuracy, check the dip frequency from both high and low sides of the resonant circuit tuning. A significant difference in frequency readout from either side indicates overcoupling between the dip adapter circuit and the resonant circuit under test. Move the adapter coil $L_1$ away from the test circuit until the dip indication is just visible. This amount of coupling should provide maximum accuracy. (If there is difficulty in finding a dip, overcouple the adapter until a dip is found, then loosen the coupling, and make a final check of frequency. Generally, the dip will be more pronounced when it is approached from the direction that causes the meter reading to rise.)

6. If there is doubt as to whether the adapter is measuring the resonant frequency of the desired circuit or some nearby circuit, ground the circuit under test. If there is no change in the adapter dip reading, the resonance of another circuit is being measured.

7. The area surrounding the circuit being measured should be free of wiring scraps, solder drippings, etc., as the resonant circuit can be affected by them (especially at higher frequencies), resulting in inaccurate frequency readings. Keep fingers and hands as far away as possible from the adapter coil (to avoid adding body capacitance to the circuit under test).

8. All other factors being equal, the nature of a dip indication provides an approximate indication of the test circuit's Q factor. Generally, a sharp dip indicates a high Q, whereas a broad dip shows a low Q.

9. The dip adapter may also be used to measure the frequency to which a resonant circuit is tuned. The procedure is essentially the same as that for presetting the resonant frequency (steps 1 through 8), except that the signal generator is tuned for a maximum dip (CB set still cold and test circuit untouched). The resonant frequency to which the test circuit is tuned is then read from the counter, or the signal generator dial if the counter is omitted. When making this test, watch for harmonics, which also produce dip indications.

## 2-13 MISCELLANEOUS CB SERVICE EQUIPMENT

There are many items of equipment that will make life easier for the CB service technician, but they are not absolutely essential for CB service. The following are some examples.

### 2-13.1 Base Station CB Set and Antenna

The uses of a known good CB set and base station (or shop) antenna are obvious. You can check operation of a suspected set on the good shop antenna. If the set performs properly with the shop antenna, but not with its own antenna, the problem is localized. You can reverse the procedure and test the suspected antenna with a known good CB set. Also, you can communicate between the shop and another station (either mobile or base station) to check operation, before and after service.

*Walkie-talkie CB.* A walkie-talkie CB may also be used for communications from the shop to remote locations for field-strength tests, etc. Keep in mind that the walkie-talkie must be licensed under Part 95 of the FCC regulations if you communicate with other CB stations (unless you communicate only between unlicensed walkie-talkies). A walkie-talkie may also be used to track down electrical interference.

### 2-13.2 Power Supply and Isolation Transformer

A well-equipped CB service shop will have at least two power supplies; one ac supply, variable from about 100 to 130 V ac; and one dc supply, variable from about 10 to 15 V dc. There are a number of commercial power supplies that meet these requirements, so we will not go into circuits.

Most commercial power supplies include a transformer. Often, this is a variable auto-transformer (or variac). The use of a transformer in the power supply eliminates the need for an *isolation transformer* as discussed in Sec. 2-1.

The power supply should have an ammeter to indicate the amount of power being drawn by the CB set being serviced. This will make it possible to detect problems in the set (such as shorts or opens) that result in an abnormal amount of power input (either too high or too low).

### 2-13.3 Distortion Meters

Some service shops include distortion meters or distortion analyzers. There are two basic types: the *intermodulation distortion analyzer*, and the *harmonic distortion analyzer*. Before you rush to buy either of these instruments, consider the following. The basic purpose of any distortion meter or analyzer is to measure the amount of distortion (usually as a percentage), not to locate the cause of distortion. The only circuits in a CB set where distortion meters may be used effectively are the audio/modulation circuits. These circuits are designed for operation at frequencies below about 3 kHz. Distortion meters are usually used in audio service work where frequencies are in the dc to 20 kHz range. Generally, if there is sufficient distortion in the audio/modulation circuits to be a problem, you will hear it in both transmission and reception.

### 2-13.4 Commercial Receivers, Spectrum Analyzers, and TV Sets

In addition to checking that a CB set produces the correct signals on all channels, it is helpful to know that the set is not producing any other signals! For example, the final RF amplifier in the transmitter may break into oscillation (if not properly neutralized) and produce signals at undetermined frequencies. This may not show up on the channel being used, or on any other channel. The problem is most common in vacuum-tube sets. More likely, the CB transmitter may produce harmonics that interfere with television and with other radio communications services. All of these extra signals (generally referred to as *spurious signals* in FCC regulations and service literature) are illegal, and certainly undesirable.

The ideal instrument for detecting undesired signals from a CB set is the *spectrum analyzer*. A laboratory spectrum analyzer is essentially a wide-band receiver with an oscilloscope output. The receiver local oscillator is swept in frequency, together with the oscilloscope horizontal sweep. Thus, the receiver is tuned (or swept) rapidly across a wide range of frequencies, and signals present on any or all of the frequencies are displayed simultaneously. Although the spectrum analyzer is ideal, it is also very expensive; it is generally restricted to laboratory use or commercial broadcast work.

You can do essentially the same job with a good *communications-type receiver*. The communications receiver should have a BFO provision as well as an S-meter. In addition to using the communications receiver for signal checks, you can monitor CB transmissions, and the receiver may also be used for WWV checks as described in Sec. 2-6.4.

Most of the interference caused by CB sets is on the television channels. A *television set* in the shop will quickly indicate whether a CB set being serviced is causing any interference. The shop TV set may help you settle some disputes concerning TV interference problems. However, before you become overconfident, keep the following in mind. Most CB interference enters the TV set through the IF amplifiers, and the IF amplifiers of all TV sets do not operate on the same frequency; some TV sets use the 22 to 28 MHz range. Other sets use the 41 to 47 MHz range, and some very old TV sets use other IF ranges. So it is possible for a CB set to produce interference on one TV set and not on another, with both TV sets located in the same room and tuned to the same channel.

## 2-14 SPECIAL TEST SETS FOR CB SERVICE

There are a number of test sets designed specifically for CB service. Some of the sets are for field use, while others are for the bench or shop. Still other sets may be used in either the shop or field. The following paragraphs describe some of these sets. Keep in mind that these are not the only test sets available, now and in the future, but represent a cross section of those that are practical for CB service. Also, the most important required functions of a test set used in CB service are described. Thus, the information may be used as a guide in selecting the right test set, or combination of sets, for your particular needs.

### 2-14.1 Power/SWR Meter

Figures 2-29 and 2-30 show a combined power/SWR test set (Archer 21-520) suitable for both amateur radio and CB. The set measures power up to 1,000 W, SWR up to 1:3, and has an impedance of 50 Ω at frequencies between 2 and 175 MHz. In use, the test set is connected between the CB set and antenna to continually monitor power output and SWR.

**Figure 2-29:** Archer 21-520 (Radio Shack) power/SWR meter

The set is calibrated for SWR when the FWD/REF switch $S_1$ is set to FWD, the transmitter is keyed, and the CAL potentiometer $VR_1$ is adjusted until the SWR meter $M_1$ is at full scale. The POWER meter $M_2$ has one scale that indicates 10, 100, or 1,000 W (full scale), depending on the position of the selector switch.

As shown in Fig. 2-30, a *directional coupler* is used for SWR measurements, as discussed in Sec. 2-11. For power measurement, part of the RF signal in the directional coupler is picked off by $C_3$, rectified by $CR_3$, and appears on POWER meter $M_2$. Variable resistors $R_5$, $R_6$, and $R_7$ permit each of the POWER meter scales to be calibrated separately. Normally, only the 10 W scale is used for CB, since an AM CB set produces only 4 W output (legally). The test set will operate with SSB, but will not provide accurate power indications, since the meter is not normally adjusted to read PEP (as discussed in Sec. 2-8).

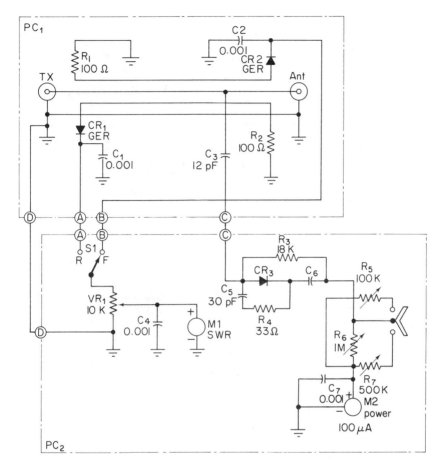

**Figure 2-30:** Archer 21-520 (Radio Shack) power /SWR meter schematic

## 2-14.2 SWR Meter/Antenna Switch

Figures 2-31 and 2-32 show a combined SWR meter and antenna switch (Archer 21-521) suitable for both amateur radio and CB. The set will operate at power outputs up to 1,000 W, measure SWR up to 1:3, and has an impedance of 50 Ω at frequencies between 2 and 30 MHz. In use, the test set is connected between the CB set and two antennas. Operation with either antenna (but not both simultaneously) is selected by the ANT 1/ANT 2 switch. The set continuously monitors SWR of the selected antenna.

**Figure 2-31:** Archer 21-521 (Radio Shack) SWR meter/antenna switch

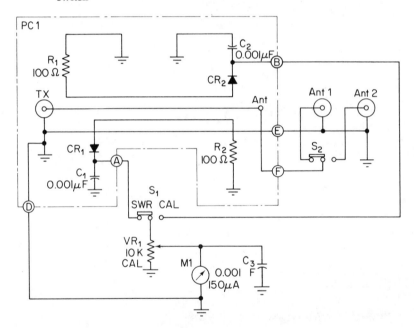

**Figure 2-32:** Archer 21-521 (Radio Shack) SWR meter/antenna switch schematic

The set is calibrated for SWR when the SWR/CAL switch $S_1$ is set to CAL, the transmitter is keyed, and the CAL potentiometer $VR_1$ is adjusted until the SWR meter $M_1$ is at full scale. As shown in Fig. 2-32, a *directional coupler* is used for SWR measurements, as discussed in Sec. 2-11.

The desired antenna is selected by ANT 1/ANT 2 switch $S_2$. The manufacturer recommends that if one antenna is not connected, even temporarily, a dummy load (Sec. 2-7) be connected at the open antenna

Sec. 2-14   Special Test Sets for CB Service    113

connector. This will prevent the transmitter from operating without a load, should the ANT switch be inadvertently set to the unused or open antenna connector.

### 2-14.3   Field Strength/SWR Meter

Figures 2-33 and 2-34 show a combined field strength and SWR meter (Archer 21-525) suitable for both amateur radio and CB. The set

**Figure 2-33:**  Archer 21-525 (Radio Shack) field strength/SWR meter

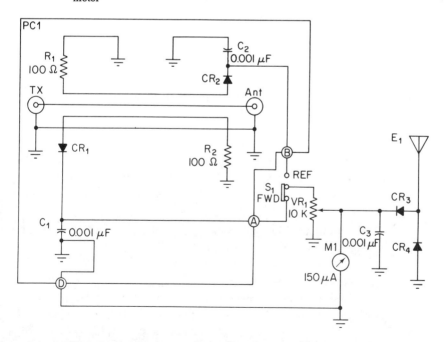

**Figure 2-34:**  Archer 21-525 (Radio Shack) field strength/SWR meter schematic

will operate at power outputs up to 1,000 W, will measure SWR up to 1:3, has an impedance of 50 Ω at frequencies between 2 and 175 MHz, and includes a removable rod-type antenna for field strength measurement. For SWR measurement, the test set is connected between the CB set and antenna and will provide continual SWR readings. For field strength operation, the removable rod antenna is connected to the test set FS ANT jack, and the CB set is operated with its own antenna.

The set is calibrated for SWR when the FWD/REF switch $S_1$ is set to FWD, the transmitter is keyed, and the CAL potentiometer $VR_1$ is adjusted until the SWR meter $M_1$ is at full scale. As shown in Fig. 2-34, a *directional coupler* is used for SWR measurements, as discussed in Sec. 2-11.

For field strength measurement, the transmitted CB signal is picked up by rod antenna $E_1$, rectified by $CR_3$ and $CR_4$, and appears on the meter $M_1$ (the same meter that is used for SWR). Keep in mind that the readout is *relative field strength*, with the meter marked off in arbitrary divisions of 1 through 5.

### 2-14.4 CB Tester

Figures 2-35 and 2-36 show a multipurpose tester (Archer 21-526) suitable for both amateur radio and CB. The set will measure power outputs up to 10 W, SWR up to 1:3, percentage of modulation between 10% and 100%, and has an impedance of 50 Ω at frequencies between 2 and 175 MHz. In use, the test set is connected between the CB set and antenna to continually monitor power output, SWR, and percentage of modulation, as selected by a front panel switch.

The same meter is used for all three functions. Operation of the meter is controlled by the selector switch as follows:

> In SWR CAL, the meter $M_1$ is connected to read the forward voltage developed across $R_2$ in the directional coupler. This voltage is rectified by $CR_1$ and applied to meter $M_1$ through SWR CAL potentiometer $VR_1$, which is adjusted until $M_1$ is at full scale.
>
> In SWR, the meter $M_1$ is connected to read the reflected or return voltage developed across $R_1$ in the directional coupler. This voltage is rectified by $CR_2$ and applied to meter $M_1$ through $VR_1$.
>
> In RF POWER, the meter $M_1$ is connected to read the forward voltage developed across $R_2$ in the directional coupler. This voltage is rectified by $CR_1$ and applied to $M_1$ through resistors $R_3$ and $R_7$. Variable resistor $R_3$ permits the meter $M_1$ to read 10 W on full scale.

**Figure 2-35:** Archer 21-526 (Radio Shack) CB tester

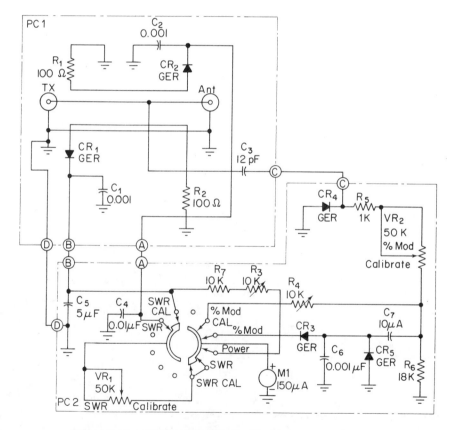

**Figure 2-36:** Archer 21-526 (Radio Shack) CB tester schematic

In MOD CAL, the meter $M_1$ is connected to read voltage tapped from the directional coupler by $C_3$. This voltage is rectified by $CR_4$ and applied to meter $M_1$ through MOD CAL potentiometer $VR_2$, which is adjusted until $M_1$ is at full scale with an *unmodulated* RF output from the transmitter. Variable resistor $R_4$ provides calibration for $M_1$ in the MOD CAL position.

In MOD, the meter $M_1$ is again connected to read voltage tapped from the directional coupler by $C_3$. The meter $M_1$ reads modulated RF output from the transmitter. The audio or modulation portion of the RF carrier is rectified by $CR_3$ and $CR_4$.

### 2-14.5 Two-way Radio Test Meter

Figures 2-37 and 2-38 show a multipurpose tester (Pace P-5425) suitable for checking transceivers in the 25 to 50 MHz range. The instrument is particularly suited for testing mobile CB units. The tester will measure power outputs up to 25 W (or 250 W when an external

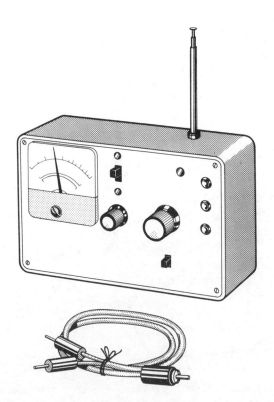

**Figure 2-37:** Pace P-5425 two-way radio test meter

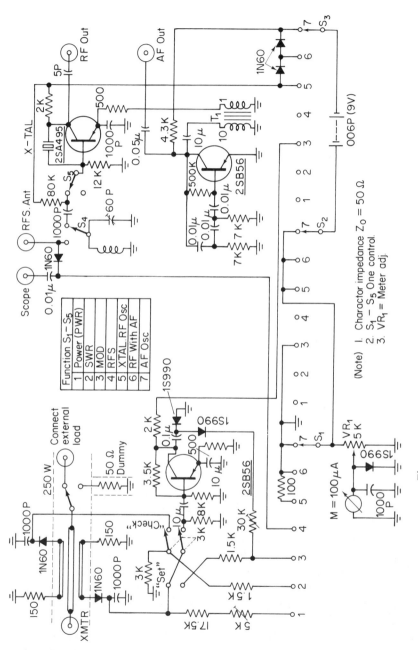

**Figure 2-38:** Pace P-5425 two-way radio test meter schematic

dummy load is used), SWR up to 1:3, percent of modulation up to 100%, relative field strength, and crystal activity (on a good-bad basis). The tester also provides a built-in 25 W dummy load, a crystal-controlled RF oscillator at 27 MHz, and an audio oscillator at 1,000 Hz (which can also be used to modulate the 27-MHz RF oscillator).

The same meter is used for all functions. Operation of the meter is controlled by the selector switches as follows:

> In PWR, the meter is connected to read the forward voltage applied to the dummy load (25 W) or to an external load (up to 250 W).
>
> In SWR, the meter is connected to read both forward and reflected voltages in the directional coupler, depending upon the position of the SET-CHECK switch.
>
> In MOD, the meter is connected to read both forward and reflected voltages in the directional coupler, depending upon the position of the SET-CHECK switch. In the CHECK position, the meter is connected through the 2SB56 transistor circuit to read only the audio or modulation voltage of the RF carrier, as a percentage of modulation.
>
> In RFS, the meter is connected to read the rectified signal present on a telescopic antenna connected to the RFS ANT jack. The rectified or detected signals are also available at the SCOPE terminal, permitting the audio or modulation signals to be displayed on an external oscilloscope.

In XTAL/RF OSC, the meter is connected to read a portion of the 9V (from the tester's internal battery) applied to the RF oscillator. A crystal to be tested is inserted into the RF oscillator XTAL socket and the meter is adjusted to read full scale by potentiometer $VR_1$. The crystal is then removed, and meter needle or pointer drops back to some point less than full scale. If the pointer stops within the GOOD zone of the XTAL scale on the meter, the crystal under test is satisfactory for use in a CB set (but not necessarily on-frequency). If a defective crystal is tested, the RF oscillator will not oscillate, and the meter pointer will remain in the BAD zone after the crystal is removed.

With a good crystal in the XTAL socket and the selector at XTAL/RF OSC, an unmodulated RF output is available from the RF OUT jack. This signal may be used to test the operation of a CB receiver, or as a frequency standard (depending upon the accuracy of the crystal).

In RF WITH AF, the meter is grounded and produces no indication. Power (9V) is applied to both the RF and AF oscillators. With a good crystal in the XTAL socket, an RF output (modulated at about 1,000 Hz)

Sec. 2-14  Special Test Sets for CB Service                                    119

is available from the RF OUT jack. A signal is also available from the AF OUT jack.

If the crystal used in the RF oscillator is at a frequency corresponding to a CB channel, the RF OUT signal may be used for *signal injection* to check operation of the CB receiver from antenna to loudspeaker. (The 1,000 Hz tone should be heard in the loudspeaker.)

In AF OSC, the meter is grounded and produces no indication. Power (9V) is applied to the AF oscillator, and an audio voltage at about 1,000 Hz is available from the AF OUT jack. This audio signal may be used for signal injection to check operation of the CB receiver audio circuits (typically from detector or volume control to loudspeaker).

### 2-14.6  Two-way Radio Tester

Figures 2-39 and 2-40 show another multipurpose tester (Pace P5430) suitable for checking transceivers in the 25 to 30 MHz range. The

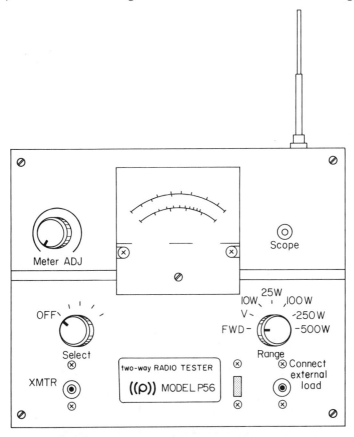

**Figure 2-39:**  Pace P-5430 two-way radio tester

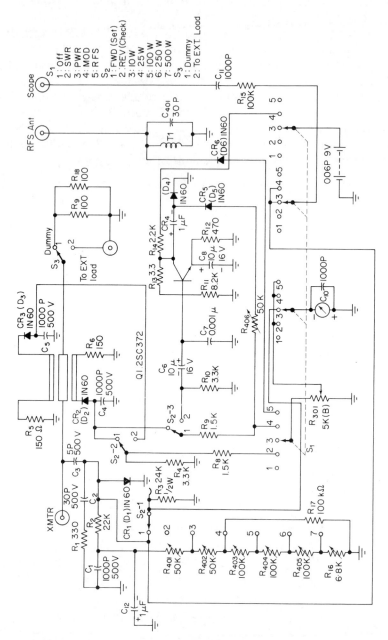

**Figure 2-40:** Pace P-5430 two-way radio tester schematic

Sec. 2-14   Special Test Sets for CB Service                                   **121**

instrument is particularly suited for test of mobile CB. Operation of this test set is similar to that of the test set described in Sec. 2-14.5. However, the test set shown in Figs. 2-39 and 2-40 will not produce an RF or AF output voltage, nor will it check crystal activity.

# 3

# TYPICAL CB CIRCUITS

This chapter describes the theory of operation for a number of CB sets and examines a cross section of present-day CB circuits. These include typical circuits for vacuum tube, AM solid-state mobile, AM solid-state base station, phase-locked loop (PLL), single-sideband (SSB), and walkie-talkie sets. By studying these circuits thoroughly, you should have no difficulty in understanding the schematic and block diagrams of any CB set. This understanding is essential for logical troubleshooting and service, no matter what type of electronic equipment is involved.

Note that no attempt has been made to duplicate the full schematics for all circuits. Such schematics are found in the service literature for the particular set. Instead of a full schematic, the circuit descriptions are supplemented with partial schematics that show such important areas as signal flow paths, input/output, adjustment controls, test points, and power source connections. As discussed in Chapters 1 and 4, these are the areas most important in troubleshooting. By reducing the schematics to these areas, you will find the circuit easier to understand, and you will be able to relate circuit operation to the corresponding circuit of the CB set you are servicing.

## 3-1 VACUUM-TUBE CB CIRCUITS

Figure 1-4 is the block diagram of the Sonar FS3023 vacuum-tube CB set. The overall description of the set is covered in Sec. 1-2 of Chapter 1 and will not be repeated here. Instead, we will go directly to the circuit descriptions.

## 3-1.1 Power Supply

Figure 3-1 shows the power supply circuits. Note that a solid-state power supply is used, even though the set is vacuum-tube. The secondary output of transformer $T_9$ is rectified by a full-wave bridge consisting of $D_4$ through $D_7$. The +285 V is filtered by $CH_1$ and $C_{82}$–$C_{84}$ and is applied to the receiver or transmitter circuits through contacts 2, 6, and 10 of switching relay $RY_1$, and directly to the "all time" B+ circuits (audio or modulator, frequency synthesizer). The "all time" circuits are required for both reception and transmission. The +140 V is filtered by $R_{77}$ and $C_{85}$–$C_{86}$ and is used to operate relay $RY_1$. When the microphone push-to-talk (PTT) switch is pressed, the $RY_1$ coil is returned to ground, contacts 6 and 10 are closed, and +285 V is applied to the transmission circuits.

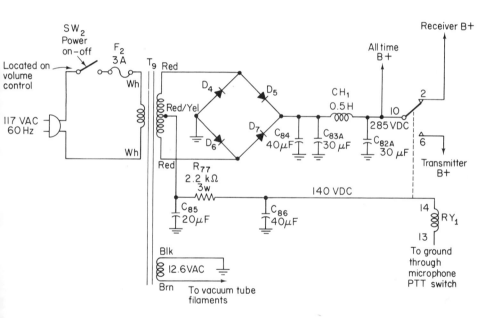

**Figure 3-1:** Vacuum-tube CB power supply circuits

When the PTT switch is released, power is removed from the $RY_1$ coil, contacts 2 and 10 are closed, and +285 V is applied to the receiver circuits.

## 3-1.2 Receiver

As shown in Fig. 3-2a, 27 MHz signals at the antenna are applied through the low-pass filter and contacts 4–12 of $RY_1$ to the input of RF amplifier $V_1$, which is neutralized by $C_4$ and tuned by $L_1$. The output of $V_1$ is tuned by $L_2$ and applied to the grid of first mixer $V_{2A}$. The cathode of

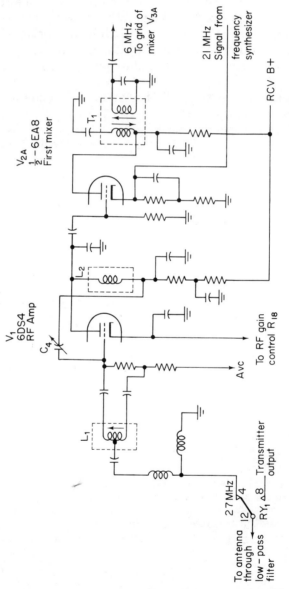

**Figure 3-2A:** Vacuum-tube CB receiver circuits (RF amp and first mixer)

Sec. 3-1   Vacuum-tube CB Circuits                                           125

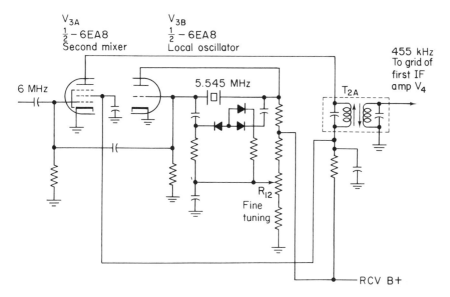

**Figure 3-2B:** Vacuum-tube CB receiver circuits (second mixer and local oscillator)

$V_{2A}$ receives a 21 MHz signal from the frequency synthesizer. The combined 27 and 21 MHz signals produce a 6 MHz signal, which is applied to the grid of second mixer $V_{3A}$ through transformer $T_1$. As shown in Fig. 3-2B, the grid of $V_{3A}$ also receives a 5.545 MHz signal from local oscillator $V_{3B}$ through $C_{14}$. Potentiometer $R_{12}$ provides for fine tuning of oscillator $V_{3B}$. The combined 6 and 5.545 MHz signals produce a 455 kHz signal, which is applied to the grid of first IF amplifier $V_4$ through both sections of $T_2$.

RF GAIN potentiometer $R_{18}$ (Fig. 3-2c) provides a variable bias for the cathodes of both $V_1$ and $V_4$. This variable bias sets the sensitivity of both the RF and IF stages so as to prevent distortion on very strong incoming signals. The output of $V_4$ is applied to the grid of second IF amplifier $V_5$ through $T_3$. In turn, the output of $V_5$ is applied to the cathode of detector $V_{6A}$. During reception, meter $M_1$ functions as an S-meter to measure signal strength. The signal appearing at the screen grid of $V_5$ is measured by $M_1$. Potentiometer $R_{56}$ provides for zero adjustment of $M_1$. (The antenna is disconnected and $R_{56}$ is adjusted for a zero on $M_1$ with normal background noise.)

The detected audio output at the plate of $V_{6A}$ (Fig. 3-2d) is applied to the AVC line through $R_{26}$ and to the cathode of noise limiter $V_{6B}$ through $R_{29}$. The AVC line provides a bias that varies with signal strength. An increase in signal strength increases the negative bias on the grids of the

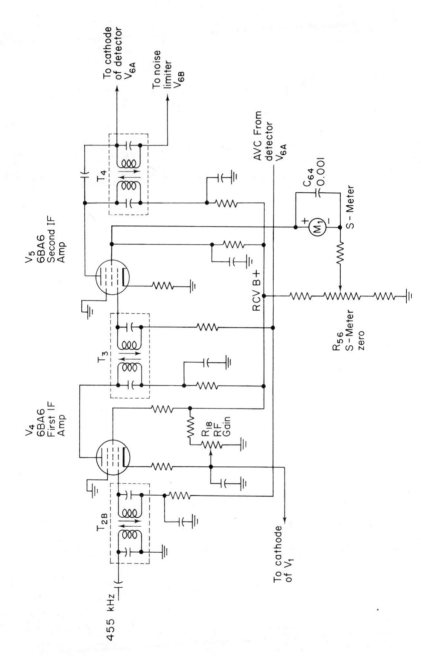

**Figure 3-2C:** Vacuum-tube CB receiver circuits (first and second IF amplifiers)

Sec. 3-1 Vacuum-tube CB Circuits

affected stages ($V_1$, $V_4$, $V_5$, and $V_{2A}$) and serves to desensitize the stages, thus offsetting the increase in signal strength. A decrease in signal strength produces the opposite effect.

The audio signals at the cathode of $V_{6B}$ are applied to the volume control $R_{31}$ through one of two paths, depending upon the setting of ANL (automatic noise limiter) switch $SW_1$. In the ON position, switch $SW_1$ is open, and the audio is applied through the cathode-plate of $V_{6B}$. Under these conditions, the ANL function is in effect. The cathode of $V_{6B}$ is biased so that $V_{6B}$ will conduct audio signals of normal level. Very large noise signals bias the cathode of $V_{6B}$ so that $V_{6B}$ is cut off, and no audio is applied to $R_{31}$. In the OFF position, switch $SW_1$ is closed, $V_{6B}$ is bypassed, and all audio and/or noise signals pass to $R_{31}$. The output of VOLUME control $R_{31}$ is applied to audio/modulation amplifier $V_{12A}$ through contacts 1–9 of $RY_1$.

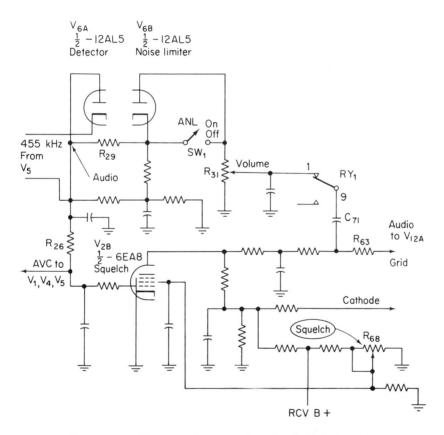

**Figure 3-2D:** Vacuum-tube CB receiver circuits (detector, noise limiter, and squelch)

The AVC line is also connected to the grid of squelch control $V_{2B}$, which receives B+ only during receive. The output of $V_{2B}$ sets the bias on the grid of audio amplifier $V_{12A}$. In the absence of a signal, $V_{2B}$ can be set by adjustment of SQUELCH control $R_{68}$ so that $V_{12A}$ is biased off. Under these conditions, the audio/modulation section is cut off and no background noise appears on the loudspeaker. When a signal is applied, the AVC line voltage changes, and $V_{2B}$ removes the cutoff bias from $V_{12A}$ to permit passage of the audio signal.

### 3-1.3 Transmitter

As shown in Fig. 3-3, vacuum tubes $V_9$, $V_{10}$, and $V_{11}$ receive B+ only during transmit. Second synthesizer mixer $V_{9A}$ receives a 21 MHz signal from the frequency synthesizer through transformer $T_5$. $V_{9A}$ also receives a 6 MHz signal from oscillator $V_{9B}$ through the coupled cathodes. Oscillator $V_{9B}$ is crystal controlled by $X_2$. Capacitor $C_{47A}$ provides for fine tuning of $X_2$. The combined 6 and 21 MHz signals produce a 27 MHz signal at the plate of $V_{9A}$.

The 27 MHz signal is applied to the grid of RF amplifier $V_{10}$ through $T_6$. The 27 MHz signal is amplified by $V_{10}$ and applied to the grid of power amplifier $V_{11}$ for further amplification. $L_4$ provides simultaneous tuning for the output of $V_{10}$ and the input of $V_{11}$. A test point is provided at the junction of $R_{52}$ and $R_{53}$ to monitor grid current of $V_{11}$. Capacitor $C_{55}$ provides neutralization for $V_{11}$. Note that the plate of $V_{11}$ receives B+ through a secondary winding of modulation transformer $T_8$. This winding (yellow-orange) is used only during transmit when $T_8$ functions as the modulation transformer. A separate winding (black-green) of $T_8$ supplies the loudspeaker during receive when $T_8$ functions as an audio output transformer.

The output of $V_{11}$ is tuned by $L_5$, $C_{62}$, and $C_{63}$, and is applied to the antenna through $L_{6A}$, contacts 8–12 of $RY_1$, and the low-pass filter. The output of $L_{6B}$ is rectified by $D_3$ and applied to meter $M_1$, which acts as a relative RF strength meter during transmit.

### 3-1.4 Audio and Modulation Section

As shown in Fig. 3-4, vacuum tubes $V_{12}$ and $V_{13}$ receive B+ during both transmit and receive. During transmit, $V_{12}$ and $V_{13}$ function as a speech amplifier and modulator. Input to $V_{12A}$ is from the audio clipper through contacts 5–9 of $RY_1$. Output from push-pull amplifier $V_{13}$ is taken from the yellow-orange secondary winding of $T_8$. As discussed in Sec. 3-1.3, this winding is between the plate of $V_{11}$ and B+ and serves to amplitude-modulate the RF carrier.

During receive, $V_{12}$ and $V_{13}$ function as an audio amplifier. Input to $V_{12A}$ is from the volume control $R_{31}$ through contacts 1–9 of $RY_1$. Output from push-pull amplifier $V_{13}$ is taken from the black-green secondary

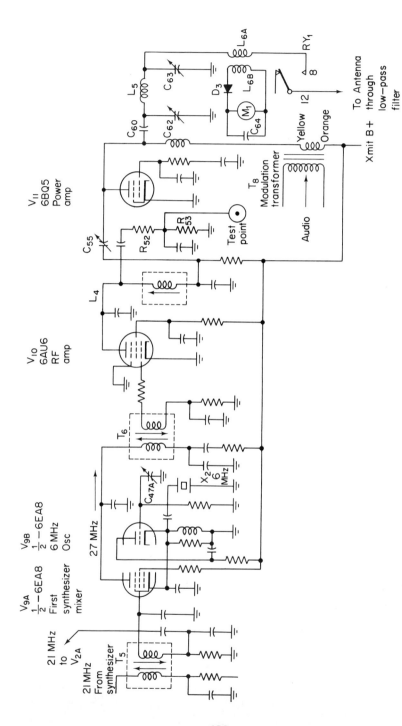

**Figure 3-3:** Vacuum-tube CB transmitter circuits

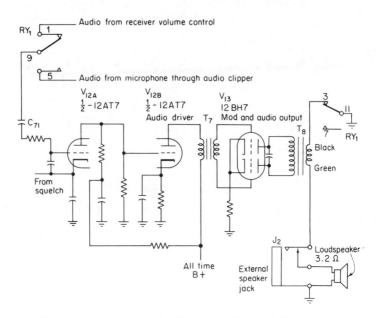

**Figure 3-4:** Vacuum-tube CB audio and modulation section

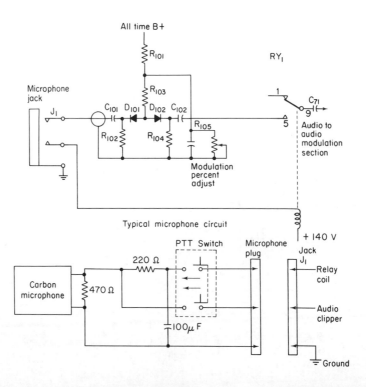

**Figure 3-5:** Vacuum-tube CB audio clipper and typical microphone circuit

Sec. 3-1 Vacuum-tube CB Circuits                                          131

winding of $T_8$. This winding is connected to the loudspeaker and to ground through contacts 3–11 of $RY_1$. During transmit, the ground is removed from the black-green winding by the opening of contacts 3–11, thus disabling the receiver.

### 3-1.5  Audio Clipper

As shown in Fig. 3-5, input to the audio clipper is from the microphone, and output is to the modulation section through contacts 5–9 of $RY_1$. Diodes $D_{101}$ and $D_{102}$ (which perform the clipping action) receive an adjustable bias from the "all-time B+" line, even though the clipper is used only during transmit. The amount of bias is determined by modulation percentage adjustment $R_{105}$, which is set to provide 100% modulation. When audio voltages developed by the microphone exceed the bias set on diodes $D_{101}$ and $D_{102}$ (plus the normal voltage drop across the diodes), the diodes conduct and limit or "clip" the voltage output to the modulation section.

### 3-1.6  Frequency Synthesizer

As shown in Fig. 3-6, vacuum tubes $V_7$ and $V_8$ receive B+ during both transmit and receive. During transmit, $V_7$ and $V_8$ provide a 21 MHz signal to the mixer $V_{9A}$ in the transmitter circuits as described in Sec. 3-1.3. During receive, $V_7$ and $V_8$ provide the 21 MHz signal to the first mixer $V_{2A}$ in the receiver circuits as described in Sec. 3-1.2. $V_7$ is a crystal-controlled oscillator producing a signal in the 16.2 to 16.45 MHz range, depending upon selection of crystals. One of six crystals is selected by the CHANNEL SELECTOR. Capacitor $C_{28}$ provides for fine tuning of oscillator $V_7$. The output of $V_7$ is also tuned by $L_3$ and applied to the grid of mixer $V_{8A}$. $V_{8B}$ is a crystal-controlled oscillator producing a signal in the 4.7 to 4.8 MHz range, depending on the selection of four crystals by the CHANNEL SELECTOR. Capacitor $C_{38A}$ provides for fine tuning of oscillator $V_{8B}$. The output of $V_{8B}$ is applied to the cathode of mixer $V_{8A}$. The combined 21 MHz signal appears at the plate of $V_{8A}$, and is applied to the transmitter or receiver circuits through $T_5$. The cathode of $V_{8A}$ is provided with a test point. During service, this test point is monitored with a frequency counter. Typically, the frequency should be equal to the CB channel frequency, less 6 MHz.

### 3-1.7  Switching Control Circuits

As shown in Fig. 3-7, switching between the transmit and receive functions is controlled by relay $RY_1$. In turn, $RY_1$ is controlled by the microphone PTT switch. When the PTT switch is not pressed, the coil of $RY_1$ is deenergized by removal of the ground. Under these conditions, contacts 4–12 connect the antenna to the receiver input, contacts 3–11 complete the loudspeaker ground connection, contacts 1–9 connect the

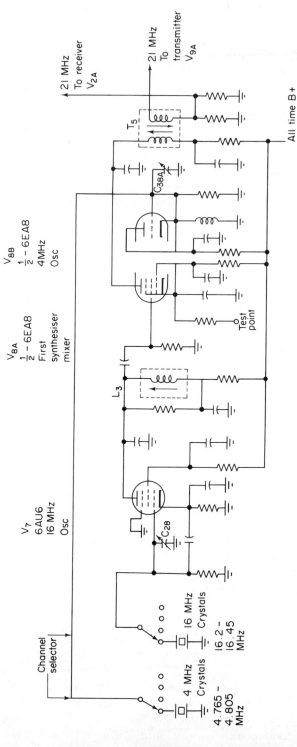

**Figure 3-6:** Vacuum-tube CB frequency synthesizer

Sec. 3-1  Vacuum-tube CB Circuits                                                133

receiver output at the volume control to the audio section input, and contacts 2–10 apply B+ to the receiver circuits. When the PTT switch is pressed, the coil of $RY_1$ is energized by completion of the ground. Under these conditions, contacts 8–12 connect the antenna to the transmitter output, contacts 3–11 are open to remove the loudspeaker ground connection, contacts 5–9 connect the audio clipper output to the modulator section input, and contacts 6–10 apply B+ to the transmitter circuits.

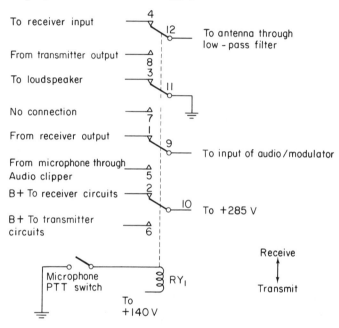

Figure 3-7: Vacuum-tube CB switching control circuits

### 3-1.8  Low-pass Filter

As shown in Fig. 3-8, the low-pass filter is a three-section filter. The inductance and capacitance values are selected to cut or attenuate signals above 30 MHz.

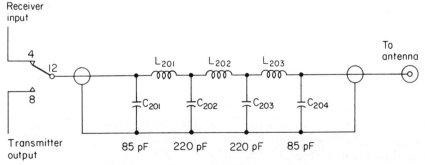

Figure 3-8: Vacuum-tube CB low-pass filter

## 3-2 AM SOLID-STATE MOBILE CB CIRCUITS

Figure 3-9 is the block diagram of the Pace CB144 solid-state CB set. The following paragraphs describe operation of the circuits.

### 3-2.1 General Description

The CB144 is a 23-channel, crystal-controlled AM transceiver. The set is all solid-state and may be operated from any standard 13.8 V dc power source. The microphone is a 500 Ω dynamic type; the loudspeaker is 8 Ω. RF power output is 4 W into 50 Ω, with 85 to 100% modulation. Receiver input sensitivity is 0.5 μV for 10dB [(S+N)/N] from a 50 Ω antenna. Audio output is 3 W, which can be applied to an external loudspeaker for a public address (PA) function. The set includes such functions as a noise blanker, ANL, RF gain control, delta tune, and an S/RF meter.

### 3-2.2 Transmitter

The transmitter is composed of two basic sections: (1) the low-level frequency generation section (frequency synthesizer), and (2) the driver, intermediate power amplifier (IPA) and power amplifier.

The synthesizer consists of two oscillators, $Q_{16}$ and $Q_{20}$. These oscillators are discussed further in Sec. 3-2.4. Master oscillator $Q_{16}$ operates at approximately 37.5 MHz, and $Q_{20}$ operates at 10.5 MHz. As shown in Fig. 3-10, the difference of the two oscillators is obtained from mixer $Q_{21}$ and passed through a bandpass filter. The output is coupled to RF driver $Q_{22}$, which operates Class AB so that a small forward bias exists with no signal and increases with drive power. The IPA $Q_{23}$ and power amplifier $Q_{24}$ are operated Class C. (There is no current flow in $Q_{23}$ or $Q_{24}$ without power applied. With power, the reverse bias increases as power increases.) Audio, taken from the push-pull output transformer $T_{17}$ secondary, is applied to the collectors of $Q_{23}$ and $Q_{24}$. When the microphone PTT switch is pressed, audio signals from the microphone are amplified by the audio amplifier (Sec. 3-2.3) and serve to modulate $Q_{23}$ and $Q_{24}$ simultaneously. The transmitter output network is applied through a 3-section pi-filter and the normally-open transmit-receive relay contacts to the antenna.

### 3-2.3 Receiver

As shown in Fig. 3-9, the receiver is a double-conversion superheterodyne. That is, two crystal-controlled oscillators ($Q_{16}$ and $Q_{17}$) are used to convert the RF signal to an IF signal. Both oscillators are changed in frequency steps to obtain 23-channel operation.

As shown in Fig. 3-11, the first mixer $Q_2$ uses high side (37.5 MHz)

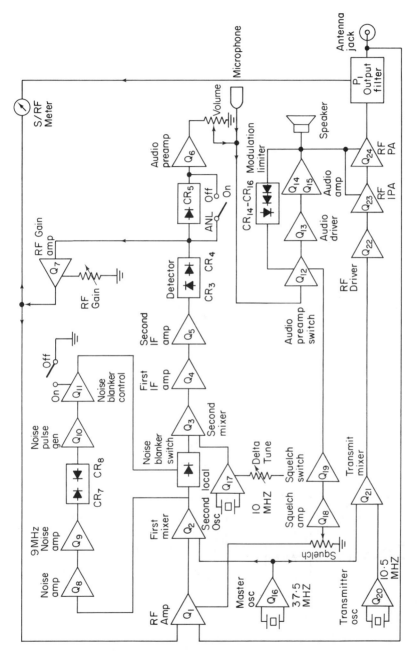

**Figure 3-9:** AM solid-state mobile CB block diagram

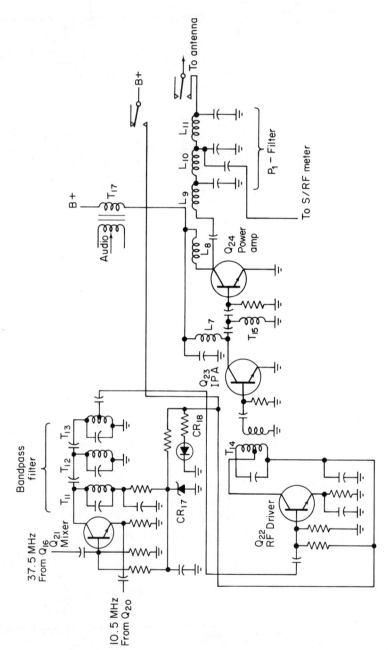

**Figure 3-10:** AM solid-state mobile CB transmitter circuits

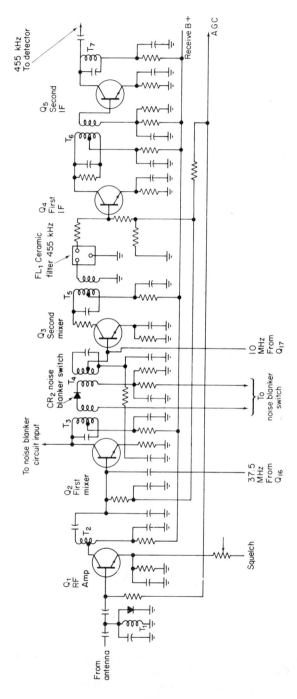

**Figure 3-11:** AM solid-state mobile CB receiver circuits

injection obtained from master oscillator $Q_{16}$ (Sec. 3-2.4), which operates during both receive and transmit. The first IF signal is then mixed with the output from the second local oscillator $Q_{17}$ (Sec. 3-2.4) at $Q_3$. The output from mixer $Q_3$ is at 455 kHz and passes through the filter circuit composed of $T_5$ and ceramic filter $FL_1$. This 455 kHz signal is amplified in two second-IF stages $Q_4$ and $Q_5$.

*Detector and noise limiter.* As shown in Fig. 3-12, the output from IF amplifier $Q_5$ is coupled through $T_7$ and $C_{23}$ to AM detector diodes $CR_3$ and $CR_4$. After detection, the audio signal passes through a noise limiter and is amplified through field-effect transistor (FET) $Q_6$. The output of $Q_6$ is applied to the arm of the volume control $R_{52}$.

Noise limiting is accomplished by the network consisting of $R_{26}$–$R_{29}$, $C_{27}$, and $CR_5$. The dc bias from the detector ($CR_3$, $CR_4$) is applied to the cathode of $CR_5$ at the junction of $R_{26}$ and $R_{27}$. The same dc bias is applied to the anode of $CR_5$ via $R_{28}$ and $R_{29}$. This forward biases $CR_5$ for normal signal amplitudes, and the audio is coupled through $CR_5$ to the gate of audio preamplifier $Q_6$.

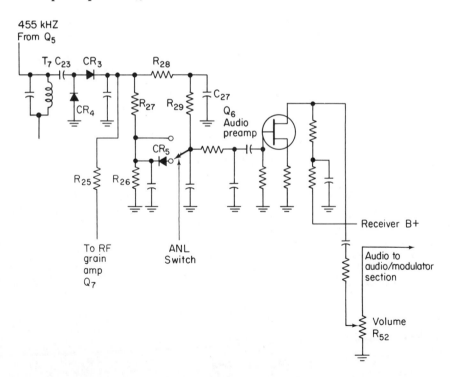

**Figure 3-12:** AM solid-state mobile CB detector and noise limiter circuit

Sec. 3-2   AM Solid-state Mobile CB Circuits                                139

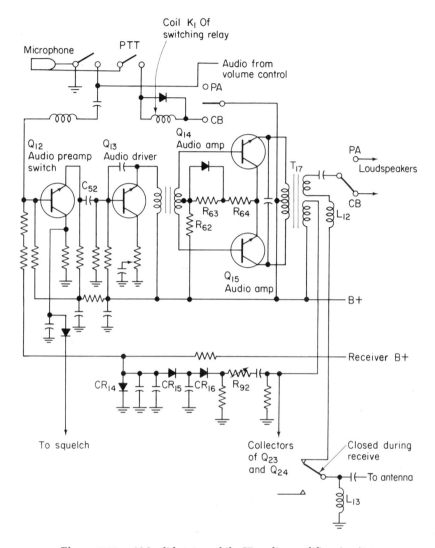

**Figure 3-13:** AM solid-state mobile CB audio amplifier circuits

When strong noise pulses are present in the signal, a higher negative bias is applied to the cathode of $CR_5$. However, the bias to the anode is not increased, because of the time constant presented by $R_{29}$ and $C_{27}$. This temporarily reverse biases $CR_5$ so that the noise pulses are clipped off. $CR_5$ is shorted out when the ANL switch is set to OFF.

*Audio amplifier.* As shown in Fig. 3-13, the audio amplifier uses an ac coupled preamp-driver followed by a common-emitter push-pull output stage. As is typical, the audio amplifier is used both during receive (to amplify the detected audio for the loudspeaker) and transmit (to amplify the microphone output for modulation of the transmitter). The audio from the collector of audio preamplifier switch $Q_{12}$ is ac coupled to driver $Q_{13}$ via $C_{52}$. (Note that $Q_{12}$ is controlled by operation of the squelch circuit.) RC combinations in the emitters of $Q_{12}$ and $Q_{13}$ boost low frequencies to compensate for losses (at those frequencies) in the transformers.

Transformer coupling is used at the input and output of the push-pull stage. Resistors $R_{62}$, $R_{63}$, and $R_{64}$ provide sufficient bias for $Q_{14}$ and $Q_{15}$ to prevent *crossover distortion*. (Transistors $Q_{14}$ and $Q_{15}$ draw a slight amount of current, even with no audio applied.) The upper winding in the secondary of $T_{17}$ couples the audio signal to the speaker (or external speaker, or PA system) during receive. This winding is returned to ground through the normally-closed transmit-receive relay contacts and chokes $L_{12}$–$L_{13}$. The lower winding couples audio (for modulation) to $Q_{23}$ and $Q_{24}$ during transmitter operation (Sec. 3-2.2).

Diodes $CR_{14}$, $CR_{15}$, and $CR_{16}$ provide for limiting of the modulation to 100%. The modulation winding of $T_{12}$ is coupled back to the input of the audio amplifier through this limiting network to provide a form of negative feedback. An increase in modulation output biases the input stage $Q_{12}$ so as to offset the increase. The amount of feedback is controlled by $R_{92}$, which is adjusted to provide a maximum of 100% modulation.

*Squelch.* As shown in Fig. 3-14, the signal level for squelch is sensed at the emitter of $Q_1$. With no signal and squelch control $R_{87}$ on full

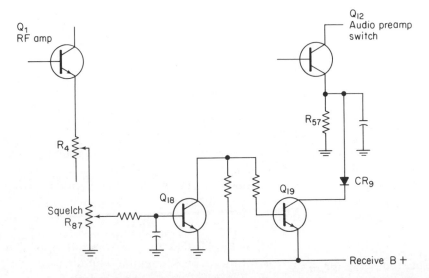

**Figure 3-14:** AM solid-state mobile CB squelch circuits

Sec. 3-2  AM Solid-state Mobile CB Circuits    141

(clockwise), there is high voltage across $R_4$ and $R_{87}$. This allows $Q_{18}$ and $Q_{19}$ to go on. When $Q_{19}$ is on, a high voltage is applied to $Q_{12}$ through $CR_9$, and $Q_{12}$ becomes reverse biased, thus allowing no audio signals through the amplifier. When an incoming signal is detected, $Q_{18}$ and $Q_{19}$ both switch off. The $Q_{12}$ emitter is grounded through $R_{57}$ in the normal manner, and audio passes through the amplifier. With squelch $R_{87}$ off, $Q_{18}$ and $Q_{19}$ are also off, and audio passes. Potentiometer $R_4$ is an internal adjustment that sets the squelch level for a given position of $R_{87}$. (Typically, $R_{87}$ is set full clockwise, and $R_4$ is adjusted so that the squelch does not open with a 1000 $\mu V$ signal at the input of $Q_1$.)

*Noise blanker.* As shown in Fig. 3-15, the first IF signal (from the collector of $Q_2$) is coupled to $Q_8$ through $C_{33}$. $Q_8$ amplifies the 10 MHz IF signal and passes it on to $Q_9$, which amplifies the signal again. The signal is passed through a 9 MHz filter and is rectified by $CR_7$ and $CR_8$. FET $Q_{10}$ reacts to noise spikes and switches on. This provides a pulse to $Q_{11}$, which turns on for an instant and thus reverse biases the noise switch $CR_2$. For that instant, no signal passes from $Q_2$ to $Q_3$, and the noise pulse is blanked.

*RF gain control and AGC.* As shown in Fig. 3-16, the front-panel RF GAIN control $R_{35}$ sets the gain of the receiver by controlling the bias to RF amplifier $Q_1$ and other stages. This system is also part of the AGC network, which reduces receiver gain when a strong signal is received. $Q_7$ monitors the dc voltage at the detector. As a signal gets stronger, the dc voltage at the anode of $CR_3$ increases and $Q_7$ turns on more. This changes the bias on $Q_1$ and other stages so as to desensitize the receiver. Since $Q_1$ is an NPN, a decrease in positive bias on the base tends to reverse bias $Q_1$.

### 3-2.4  Oscillators

As shown in Fig. 3-17, three separate oscillators are used with a total of 14 crystals. The crystals are combined in a synthesis circuit to obtain all 23 CB channels.

*Master oscillator* $Q_{16}$ is a crystal-controlled tuned-collector oscillator. Six crystals coupled to the base of $Q_{16}$ are in the frequency range of 37.6 to 37.85 MHz. A different crystal is selected for each channel. $Q_{16}$ is active in both transmit and receive. The output, taken from the secondary of $T_9$, is coupled to the base of receiver mixer $Q_2$ via $C_{10}$ and to the base of transmitter mixer $Q_{21}$ via $C_{77}$.

*Oscillator* $Q_{17}$ is a crystal-controlled RC oscillator. Four crystals, coupled to the base of $Q_{17}$, are in the frequency range of 10.140 to 10.180 MHz. A different crystal is selected for each channel. The output is taken from the emitter of $Q_{17}$ and coupled to receiver second mixer $Q_3$. This

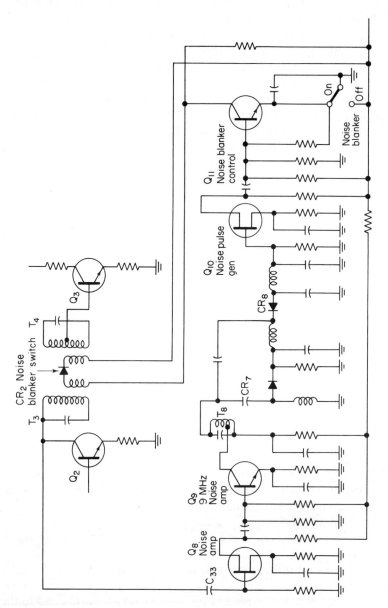

**Figure 3-15:** AM solid-state mobile CB noise blanker circuits

### Sec. 3-2  AM Solid-state Mobile CB Circuits

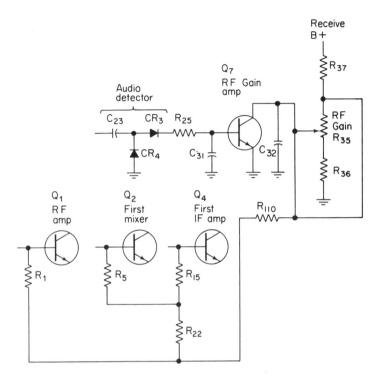

**Figure 3-16:** AM solid-state mobile CB RF gain control and AGC

frequency is then mixed with the output from the first IF amplifier $Q_2$ to obtain the 455 kHz IF. $Q_{17}$ is activated in the receive mode only. Also incorporated in the $Q_{17}$ oscillator circuit is a varactor diode $CR_{21}$ *Delta tune* circuit. Adjustment of the DELTA TUNE control $R_{81}$ varies the voltage (and thus the capacity) on $CR_{21}$. The change in $CR_{21}$ capacity shifts the frequency of $Q_{17}$ by about ±1.5 kHz. This permits reception of signals that are not exactly on the selected CB channel frequency.

*Transmitter oscillator* $Q_{20}$ is a crystal-controlled, tuned-collector oscillator. Four crystals, coupled to the base of $Q_{20}$, are in the frequency range of 10.595 to 10.635 MHz. A different crystal is selected for each channel. $Q_{20}$ is used only in the transmit mode. The output of $Q_{20}$ is taken from the secondary of $T_{10}$ and coupled via $C_{81}$ to the emitter of mixer $Q_{21}$, where the output is combined with the signal from $Q_{16}$.

#### 3-2.5 Transmit-receive Switching System

As shown in Fig. 3-18, the transmit-receive switching system is relay-controlled using two sets of contacts. These contacts are energized (move to the normally-open position) when the PTT switch is pressed.

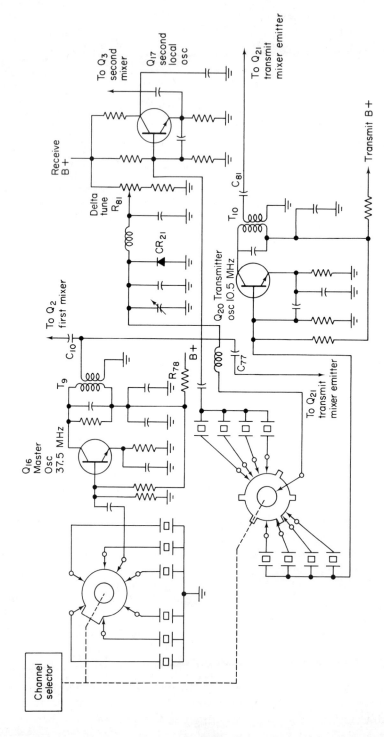

**Figure 3-17:** AM solid-state mobile CB oscillator circuits

## Sec. 3-2 AM Solid-state Mobile CB Circuits

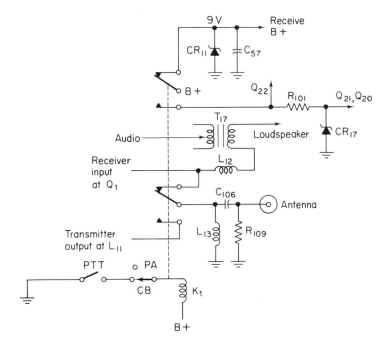

**Figure 3-18:** AM solid-state mobile CB transmit-receive switching system

B+ is continuously applied to the intermediate and power amplifiers ($Q_{23}$–$Q_{24}$), to the master oscillator $Q_{16}$ through $R_{78}$, and to the audio amplifier ($Q_{12}$, $Q_{13}$, $Q_{14}$, $Q_{15}$). Although $Q_{23}$ and $Q_{24}$ are used only during transmit, they draw no power in the absence of a drive signal. The remaining transistors with continuous B+ ($Q_{12}$ through $Q_{16}$) are used in both transmit and receive.

When the PTT switch is in the normal (receive) position, the ground return for relay coil $K_1$ is open, and the antenna is connected to the receiver RF amplifier $Q_1$ through one set of normally-closed relay contacts. At the same time, +9 V, regulated by zener diode $CR_{11}$, is supplied to all receiver circuits, including the second local oscillator $Q_{17}$.

When the PTT switch is pressed (with the PA/CB switch in CB position), the ground return for relay coil $K_1$ is closed, the relay is energized, B+ is removed from the receiver transistors, the antenna is transferred from the receiver input to the transmitter output (at $L_{11}$), and B+ is applied to transmitter transistors $Q_{20}$, $Q_{21}$, and $Q_{22}$. The voltage to $Q_{20}$ and $Q_{21}$ is regulated by zener diode $CR_{17}$ to +9 V. When the antenna is switched from the receiver to the transmitter, the ground is removed from the upper secondary output of $T_{17}$, thus rendering the loudspeaker inoperative. (This secondary winding of $T_{17}$ is normally returned to ground through the normally-closed contacts, and chokes $L_{12}$–$L_{13}$.) The microphone output is connected to the input of $Q_{12}$, and the output of the audio amplifier is applied through the lower secondary winding of $T_{17}$ to modulate the transmitter.

When the PA/CB switch is in the PA position (Fig. 3-13), the upper secondary output of $T_{17}$ is switched from the internal loudspeaker to the external public address loudspeaker jack. Also, relay coil $K_1$ is disconnected from a source of voltage (+12 V), and the relay contacts remain in the "receive" condition (the normally-closed contacts remain closed). When the PTT switch is pressed, the microphone output is connected to the input of $Q_{12}$, and the output of the audio amplifier is applied through the upper secondary winding to the public address loudspeaker.

### 3-2.6 Meter Circuitry

As shown in Fig. 3-19, the S/RF meter provides relative indications of both incoming signal strength (S-meter function) and transmitted power (RF output meter function).

In the receive mode, a dc voltage at the output of the detector $CR_3$–$CR_4$, proportional to the strength of the incoming signal, is amplified by $Q_7$ and filtered by $C_{32}$. Note that $Q_7$ is connected to B+ through the meter (in parallel with $R_{71}$), $CR_{12}$, $R_{69}$, $R_{70}$, and $R_{72}$. Potentiometer $R_{72}$ sets the level of the B+ and thus provides for adjustment of the meter.

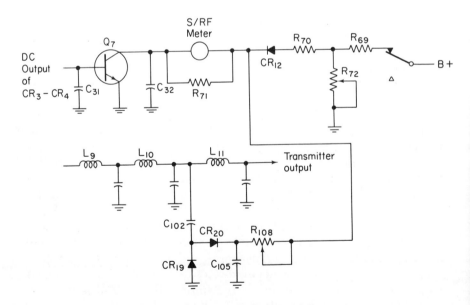

**Figure 3-19:** AM solid-state mobile CB meter circuits

Sec. 3-3 AM Solid-state Base Station CB Circuits

Typically, $R_{72}$ is set to provide an indication of "S9" with a 100 $\mu$V modulated RF signal at the receiver input.

In the transmit mode, power output signals are coupled through $C_{102}$ to diodes $CR_{19}$–$CR_{20}$, where the RF is converted to a dc voltage that is filtered by $C_{105}$. This voltage, proportional to the strength of the RF output from the transmitter to the antenna, is fed through $R_{108}$ to the meter. Potentiometer $R_{108}$ sets the amount of voltage thus applied and provides for adjustment of the meter during transmit. Typically, $R_{108}$ is set to provide a 2/3 of full scale indication on the meter when the transmitter output is 4 W.

## 3-3 AM SOLID-STATE BASE STATION CB CIRCUITS

Figure 3-20 is the block diagram of the Pace CB DX2300B solid-state CB set. The following paragraphs describe operation of the circuits.

### 3-3.1 General Description

The CB DX2300B is a 23-channel, crystal-controlled AM transceiver. The set is all solid-state and may be operated from 117 V ac or 220 V ac line power. The microphone is a high-impedance ceramic type; the loudspeaker is 3.2 $\Omega$. RF power output is 4 W into 50 $\Omega$, with 85 to 100% modulation. Receiver input sensitivity is 0.35 $\mu$V for 10dB [(S+N)/N] from a 50 $\Omega$ antenna. Audio output is 2.25 W, which can be applied to an external loudspeaker for a public address (PA) function. The set includes such functions as noise limiter, ANL, AGC, delta tune, and an S/RF meter, in addition to the usual channel selector and volume control.

### 3-3.2 Transmitter

The transmitter is composed of two basic sections: the low-level frequency generation section contained on the synthesizer output board (SOB); and the driver, intermediate power amplifier (IPA), and power amplifier located on the main printed circuit board. The SOB is connected to the main circuit board with a small coaxial cable, and thus it may be checked or serviced as a separate unit quite readily.

The SOB includes two oscillators, $Q_{14}$ and $Q_{15}$. Oscillator $Q_{14}$ operates at approximately 8 MHz, and $Q_{15}$ operates at 35 MHz. The difference of the two oscillators (approximately 27 MHz) is obtained from mixer $Q_{16}$ and passed through a bandpass filter-amplifier $Q_{17}$. The output from $Q_{17}$ is coupled by coaxial cable to driver $Q_{18}$, which operates Class AB (a small forward bias exists with no signal and increases with drive power).

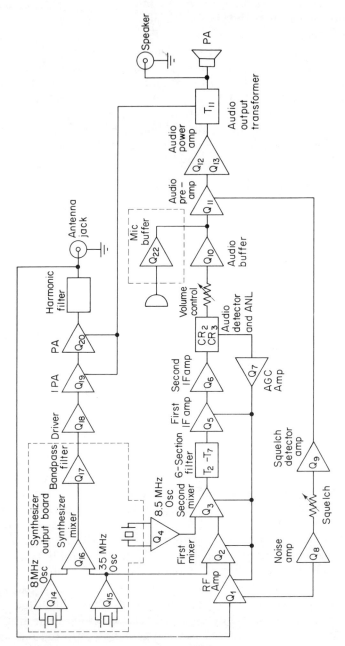

**Figure 3-20:** AM solid-state base station CB block diagram

Sec. 3-3 AM Solid-state Base Station CB Circuits 149

The IPA $Q_{19}$ and power amplifier $Q_{20}$ are operated Class C (the more drive applied, the more reverse biased their base-emitters become; no current is drawn in the absence of a drive signal). The transmitter output network is a three-section pi-filter used to eliminate harmonics that might interfere with other communications services (such as TV, business radio, etc.).

### 3-3.3 Receiver

The receiver is a double conversion superheterodyne. Both oscillators ($Q_4$ and $Q_{15}$) are crystal-controlled, and both are changed in frequency steps to obtain 23-channel operation. The first mixer, $Q_2$, uses high side injection obtained from $Q_{15}$. (Oscillator $Q_{15}$ works during both transmit and receive operations.) The second mixer $Q_3$ obtains injection from oscillator $Q_4$. The output of $Q_3$ is at 455 kHz and passes through a six-stage filter. The 455 kHz signal is further amplified by $Q_5$–$Q_6$, and is detected by $CR_2$–$CR_3$.

*Automatic gain control.* The output of the receiver detector contains the rectified audio and a dc component proportional to the carrier. As shown in Fig. 3-21, the dc component is applied to the base of AGC

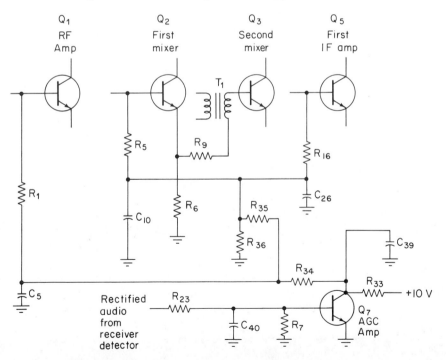

**Figure 3-21:** AM solid-state base station CB AGC circuits

amplifier $Q_7$ through the filter network $R_{23}$, $R_7$, and $C_{40}$. This positive voltage turns $Q_7$ on, causing the collector to go toward ground. $Q_1$, $Q_2$, $Q_3$, and $Q_5$ receive base bias from the collector of $Q_7$. The negative-going collector voltage reduces the bias and, consequently, the gain of these stages. At maximum signal level, the negative bias from $Q_7$ turns $Q_1$, $Q_2$, $Q_3$, and $Q_5$ completely off.

AGC for the second mixer $Q_3$ is obtained from the emitter of $Q_2$. As $Q_2$ emitter voltage is reduced by the negative-going AGC signal from $Q_7$, $Q_3$ is cut off before $Q_1$, $Q_2$, or $Q_5$. This reduces mixer noise more quickly with small increases in signal.

*PA/LCL/DIST switch.* As shown in Fig. 3-22, this switch selects PA (public address), LCL (local signal), or DIST (distant signal) functions. The receiver is at maximum sensitivity, to improve reception of weak signals, when the switch is in DIST. In the LCL position, receiver sensitivity is reduced, thus reducing most extraneous noises (as well as skip-interference). The LCL position is most useful when receiving very strong signals. The PA position disables the receiver and provides a

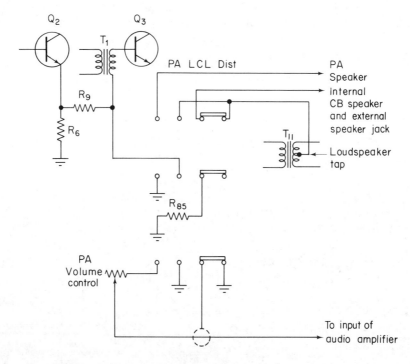

**Figure 3-22:** AM solid-state base station CB PA/LCL/DIST switch circuits

Sec. 3-3  AM Solid-state Base Station CB Circuits                               151

public address function (from the microphone, through the audio amplifier, to an external PA speaker jack).

In DIST, the base of $Q_3$ is connected to the emitter of $Q_2$, and the receiver is at maximum sensitivity. The PA volume control is disconnected from ground. The loudspeaker tap on audio amplifier output transformer $T_{11}$ is connected to the internal speaker and to an external speaker jack, but not to the PA speaker jack.

In LCL, the base current of $Q_3$ is shunted off through $R_{85}$, thus reducing the gain of $Q_3$. The receiver output drops approximately 15 dB in the LCL position. Again, the PA volume control and PA speaker jack are omitted from the circuit.

In PA, the base of $Q_3$ is shorted to ground, disabling the receiver. The PA volume control is returned to ground. The loudspeaker tap on $T_{11}$ is connected to the PA speaker jack.

*Noise limiter.* As shown in Fig. 3-23, noise limiting is accomplished with the network consisting of $R_{26}$, $R_{24}$, $R_{28}$, $C_{35}$, and $CR_4$. Direct current bias from the detector is applied to the cathode of $CR_4$ at the junction of divider $R_{26}$–$R_{24}$. The dc bias is also applied to the anode of $CR_4$ via $R_{27}$–$R_{28}$. This arrangement forward biases $CR_4$ for normal signal amplitudes, and audio is passed through $CR_4$ to the volume control.

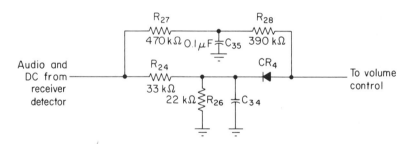

**Figure 3-23:** AM solid-state base station CB noise limiter circuits

When noise pulses are present in the signal, a higher positive bias is applied to the cathode of $CR_4$. However, the bias to the anode is not increased, because of the time constant presented by $R_{27}$ and $C_{35}$. This reverse biases $CR_4$ momentarily so that the noise pulses are clipped off. Noise pulses are usually equal to three or four times the normal 100% modulated audio signal level. Thus, noise pulses will gate $CR_4$ off, whereas a modest increase in signal level will have no effect on $CR_4$, and audio will pass. The clipping level is fixed at about 65%.

*Audio amplifier.* As shown in Fig. 3-24, the audio system is an all direct-coupled, single-ended amplifier. The input stage $Q_{10}$ is used

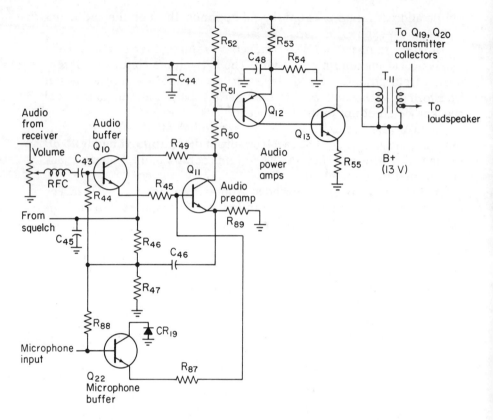

**Figure 3-24:** AM solid-state base station CB audio amplifier circuits

during receive. $Q_{10}$ is biased by a divider string in the collector of $Q_{11}$, thus providing negative dc feedback around $Q_{10}$ and $Q_{11}$. This stabilized dc voltage drives $Q_{12}$. Resistor $R_{53}$ in the emitter of $Q_{12}$ is selected to compensate for the difference in the beta of $Q_{13}$. $R_{53}$ is selected to provide about 1.25 A of current during transmit and/or PA functions. $R_{54}$ is then selected to produce about 750 mA during receive.

Input stage $Q_{22}$ is used during transmit and/or PA functions. $Q_{22}$ acts as a buffer between the microphone input and audio amplifier to prevent loading of the audio stage (in receive mode) when the set is used with a microphone that grounds the audio input (low-impedance type).

Transformer coupling is used at the output of $Q_{13}$. The secondary of $T_{11}$ couples the audio signal to the speakers (during receive or PA) or to the collectors of $Q_{19}$ and $Q_{20}$ (during transmit). Note that audio is taken from the tap of $T_{11}$ during receive and PA, and from the full secondary winding during transmit. The function of $T_{11}$ in the transmit-receive switching system is discussed further in Sec. 3-3.4.

Sec. 3-3   AM Solid-state Base Station CB Circuits                        153

*Squelch amplifier.* As shown in Fig. 3-25, the squelch sensing voltage is taken from the collector of the RF amplifier $Q_1$. With increasing signal strength, the base of $Q_1$ moves toward ground, cutting off $Q_1$. Because of dc load resistor $R_4$, the voltage at the collector of $Q_1$ increases in a positive direction when $Q_1$ is cut off. Since $Q_8$ is a PNP transistor, this positive-going voltage cuts $Q_8$ off. The cutoff or "squelch" point is determined by $R_{38}$, the squelch control. As the control is moved toward the +10 V supply, $Q_1$ must be more nearly cut off (requiring a stronger

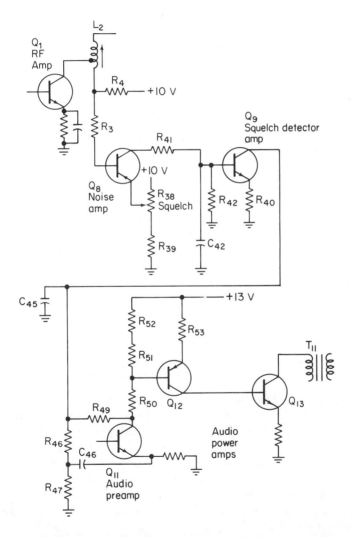

**Figure 3-25:** AM solid-state base station CB squelch amplifier circuits

signal) to cut off $Q_8$. With $Q_8$ cut off, no current flows in $R_{41}$–$R_{42}$, and this in turn cuts off $Q_9$.

Squelch operation is obtained by "starving" the divider string in the collector of $Q_{11}$. The squelch transistor $Q_9$ draws current through a 10 k$\Omega$ resistor $R_{49}$. This prevents any base current from reaching $Q_{10}$, turning $Q_{10}$ off. In turn, this shuts off $Q_{11}$. The only current then flowing through $R_{50}$, $R_{51}$, and $R_{52}$ is the small amount flowing through $R_{49}$ to $Q_9$. This current is insufficient to turn $Q_{12}$ on, and $Q_{13}$ will consequently remain off.

S-meter. As shown in Fig. 3-26, the S-meter $M_1$ is in a bridge circuit, one-half of which is formed by the collector circuit of $Q_7$, the AGC amplifier. With no signal, the voltage across the S-meter is nearly zero. With increasing signal strength, $Q_7$ conducts, causing the voltage at its collector to go negative (toward ground) with respect to the junction of $R_{31}$–$R_{32}$. This causes current flow through the S-meter. Potentiometer $R_{32}$ provides for adjustment of the S-meter. Typically, $R_{32}$ is set to provide an indication of "S9" with a 100 $\mu$V modulated RF signal at the receiver input.

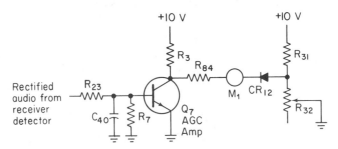

**Figure 3-26:** AM solid-state base station CB S-meter circuits

### 3-3.4 Transmit-receive Switching System

As shown in Fig. 3-27, the transmit-receive switching system is semi-electronic. That is, all of the normal transmit-receive switching functions (antenna switched between receiver input and transmitter output, audio input switched between receiver or microphone, audio output switched between loudspeaker and modulation, receiver/transmitter enabled and disabled) are controlled by one set of single-pole, double-throw relay contacts. The relay contacts are shown in the receive position, with the PTT switch shown in the normal (not pressed) position, on Fig. 3-27.

When the PTT switch is normal (receive), the antenna is connected through $C_1$ and $C_{85}$ to the receiver RF stage. Chokes $RFC_7$ and $RFC_8$

Sec. 3-3   AM Solid-state Base Station CB Circuits                          155

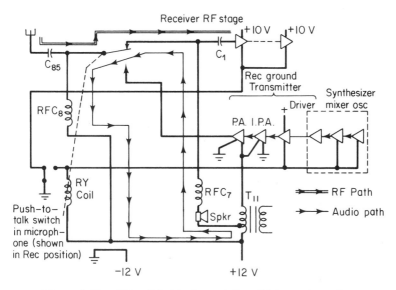

**Figure 3-27:** AM solid-state base station CB transmit-receive switching system

present a high RF impedance so that no antenna currents will flow into the audio circuit. The speaker is connected across $T_{11}$, since the chokes are essentially zero impedance at audio frequencies. Loss of audio through the antenna or RF stage is prevented by $C_1$ and $C_{85}$, which act as high impedances at audio frequencies.

When the PTT switch is pressed (transmit) the ground side of the receiver is opened, thereby disabling the receiver. The relay coil, synthesizer mixer $Q_{16}$, oscillator $Q_{14}$, and driver $Q_{18}$ are energized by completing their ground returns, the speaker is disconnected, and the transmitter is connected to the antenna.

Note that +12 V is applied to the transmitter IPA and PA stages at all times (through $T_{11}$). However, except during transmit, these stages have no drive, and thus draw no current. Receiver audio also appears on the IPA and PA collectors. However, since the stages are drawing no current, they appear as a small capacity shunting $T_{11}$.

### 3-3.5   Oscillators

As shown in Fig. 3-28, three separate oscillators are used with a total of 14 crystals. The crystals are combined in a synthesis circuit to obtain all 23 CB channels.

*Oscillator* $Q_{15}$ is a crystal-controlled Colpitts oscillator. Six crystals coupled to the base of $Q_{15}$ are in the frequency range of 34.971 to 35.221 MHz. A different crystal is selected for each channel. $Q_{15}$ is active in both

## Ch. 3 TYPICAL CB CIRCUITS

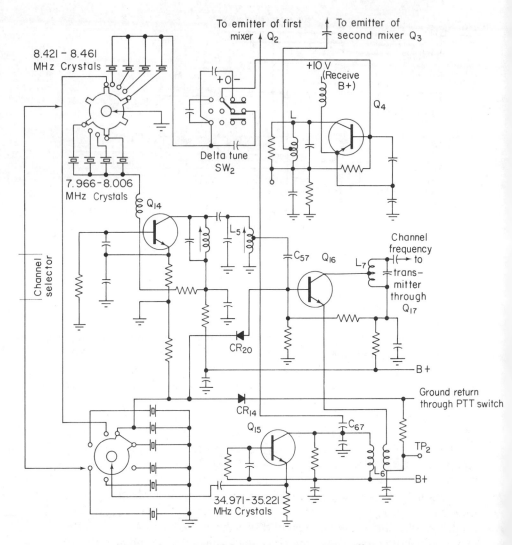

**Figure 3-28:** AM solid-state base station CB oscillator circuits

transmit and receive. The output, taken from the secondary winding of $L_6$, is connected to the emitter of transmitter mixer $Q_{16}$. Coupling to the emitter of first receiver mixer $Q_2$ is taken from the primary of $L_6$ through $C_{67}$.

Oscillator $Q_4$ is a crystal-controlled modified Hartley oscillator. Four crystals coupled to the base of $Q_4$ are in the frequency range of 8.421 to 8.461 MHz. A different crystal is selected for each channel. The output is taken from the collector of $Q_4$ and coupled to the emitter of the receiver second mixer $Q_3$. This frequency is then mixed with the output from the

Sec. 3-3    AM Solid-state Base Station CB Circuits                              157

first mixer $Q_2$ to obtain the 455 kHz IF. $Q_4$ is activated in receive only.

Oscillator $Q_{14}$ is a tuned-collector crystal oscillator. Four crystals coupled to the base of $Q_{14}$ are in the frequency range of 7.966 to 8.006 MHz. A different crystal is selected for each channel. $Q_{14}$ is only active in transmit (when the PTT switch is pressed to complete the emitter circuit). The output taken from the center tap of $L_5$ is coupled via $C_{57}$ to the base of transmitter mixer $Q_{16}$. This frequency mixed with that from $Q_{15}$ produces the channel frequency.

### 3-3.6 Power Supply

As shown in Fig. 3-29, the power supply is a constant voltage-regulated dc supply. The 117 V ac input is stepped down by power transformer $T_{10}$, rectified by bridge circuit $CR_{15}$ through $CR_{18}$, and filtered by $C_{103}$.

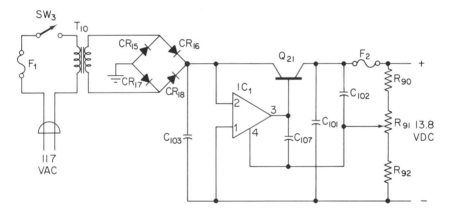

**Figure 3-29:** AM solid-state base station CB power supply circuits

The voltage is regulated through operation of IC amplifier $IC_1$ and series pass transistor $Q_{21}$, in a conventional series-feedback circuit. Feedback action alters the voltage drop across $Q_{21}$ so as to keep the dc output voltage constant despite load or line changes. Comparison amplifier $IC_1$ monitors the difference between the dc voltage at terminal 2 and a sampling of the output voltage taken from the tap of $R_{91}$ (at terminal 4 of $IC_1$). If these voltages are unequal, an error signal is produced at the output of $IC_1$ (terminal 3) that changes the conduction of $Q_{21}$. This changes the dc output voltage in opposition to the change that produced the unbalance. That is, if the dc output rises, the $Q_{21}$ conduction increases, thus lowering the dc output voltage so as to maintain a constant output. Typically, $R_{91}$ is adjusted for a 13.8 V output across $R_{90}$, $R_{91}$, and $R_{92}$.

## 3-4 PHASE-LOCKED LOOP (PLL) CB CIRCUITS

Many present-day CB sets use a PLL circuit in place of the conventional frequency synthesizer. There are two definite advantages to PLL. First, the oscillator is not only crystal-controlled, but is locked in frequency to the crystal by a feedback circuit that detects even a slight shift in phase between oscillator and crystal. This maintains frequency accuracy far better than is possible with a conventional crystal oscillator, even with well-designed circuits, in spite of power supply and temperature changes. Second, only two or three crystals are required for PLL (to cover all 23 or 40 channels), instead of the usual 12 to 14 crystals for a conventional synthesizer.

Figure 3-30 is the basic block diagram of a typical PLL system. The voltage-controlled oscillator (VCO) produces a signal at the desired frequency. In some systems, the VCO signal is at the channel frequency and is applied directly to the transmitter amplifiers, thus substituting for the signal usually obtained from a conventional frequency synthesizer.

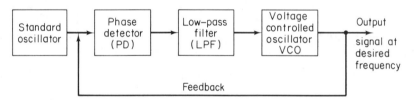

**Figure 3-30:** Basic phase-locked loop (PLL) system

In different systems, the VCO output is mixed with other signals to produce the transmitter frequency and receiver local oscillator frequencies. With any system, the VCO output is fed back and compared with a standard signal as to frequency and phase. Likewise, with any system, the VCO frequency is determined by a voltage from the phase detector (PD) applied through a low-pass filter (LPF). In most systems, the PD produces a pulsed output, which is converted to a voltage by the LPF.

The PD pulse width is determined by the frequency/phase difference of the standard oscillator and the VCO. If both oscillators are locked in exact frequency and phase, the PD pulse width is zero (or at some fixed width, depending upon the system), the LPF output is zero (or fixed), and the VCO output frequency remains fixed. If the VCO frequency deviates from that of the standard oscillator (even a fraction of one cycle), the PD pulse width changes, as does the LPF voltage to the VCO, and the VCO frequency also changes.

This change opposes the initial change in frequency. For example, if

Sec. 3-4   Phase-locked Loop (PLL) CB Circuits                                   159

the VCO output frequency is lowered from that of the standard oscillator (perhaps due to a change in temperature, power supply voltage, etc.), the PD senses this change and produces a wider pulse. In turn, the wider pulse produces an increase in voltage from the LPF and increases the VCO frequency so as to offset the undesired deviation from that of the standard. Note that in most CB set PLL systems, the PD and standard oscillator (except for the crystal) are contained in an IC, whereas the LPF and VCO are usually separate circuits.

### 3-4.1   Typical PLL System

Figure 3-31 is the block diagram of the PLL system used in the Panasonic RJ-3100. Figure 3-32 shows a partial schematic of the same circuits. In this system, the VCO output frequency (fo) becomes equal to the transmitting frequency (fT) during transmit and is equal to the receiver local oscillator frequency (fL) during receive.

Note that the IC contains two dividers in addition to the phase detector and standard oscillator. The fixed divider divides the standard oscillator frequency (fs) by 1/1024 to convert the 10.240 MHz crystal frequency (fv) to 10 kHz. The programmable divider is controlled by the channel selector and divides the mixer output frequency (fm) as necessary to produce 10 kHz.

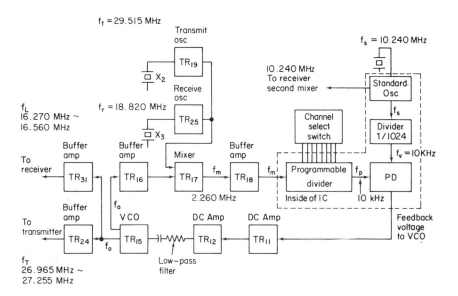

**Figure 3-31:**   PLL system block diagram

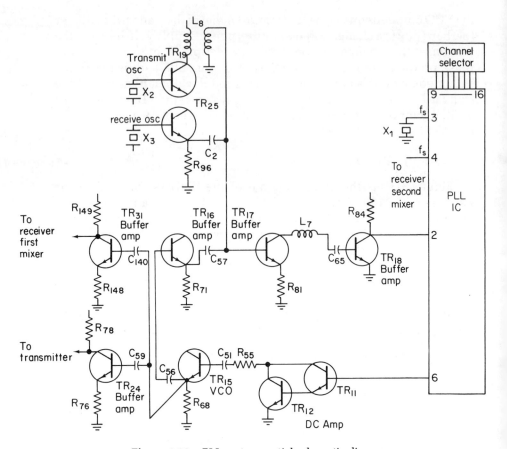

**Figure 3-32:** PLL system partial schematic diagram

For example, if channel 23 is selected, the desired VCO output must be 27.255 MHz during transmit. The VCO signal is mixed with 29.515 MHz signal from the transmitter oscillator (ft) to produce a difference frequency of 2.260 MHz. With the channel selector set at 23, the programmable divider divides the difference frequency of 2.260 MHz by 1/226 to produce 10 kHz. Note that the standard frequency (fs) of 10.240 MHz is used directly for the receiver second local oscillator signal to produce the desired 455 kHz IF. Using channel 23 again, the VCO produces 16.560 MHz, which combines in the receiver first mixer with the incoming 27.255 MHz to produce an IF of 10.695. This is combined in the receiver second mixer with the 10.240 MHz standard frequency to produce 455 kHz. In a similar fashion, all 23 channel frequencies (including the transmitter frequency and both receiver oscillator frequencies) are produced with only three crystals.

Sec. 3-4 Phase-locked Loop (PLL) CB Circuits          161

### 3-4.2 Typical PLL CB Set Circuits

Figure 3-33 is the complete block diagram for the Panasonic RJ-3100. The following paragraphs describe some of the circuits found in the RJ-3200, a companion set to the RJ-3100. Note that except for replacement of the conventional frequency synthesizer with a PLL system, most of the RJ-3100 circuits are quite similar to those of any solid-state CB set.

### 3-4.3 RF Output Filters

As shown in Fig. 3-34, these filters consist of the matching network, the low-pass filter, and the second and third harmonic traps.

The matching network $L_{16}$–$C_{137}$ converts (matches) the 23 Ω output of the RF final amplifier stage to 50 Ω, which is the impedance of the low-pass filter and antenna. The low-pass filter $L_{17}$, $L_{18}$, $C_{64}$, $C_{65}$, $C_{71}$ suppresses the second and other higher-frequency harmonics. (The manufacturer recommends that $L_{17}$–$L_{18}$ be adjusted while viewing the transmitter output on a spectrum analyzer.) The level of the harmonics must be at least 50 dB or 60 dB below the level of the fundamental signal.

The 2nd ($C_{66}$–$L_{21}$) and third ($C_{73}$–$L_{19}$) traps provide additional suppression for the second and third harmonics, respectively. These harmonics are in the 54 and 108 MHz ranges and can interfere with television and FM broadcasts.

Note that the transmitter output and receiver input are connected directly to the output filter circuit, not through a switching relay. As discussed in Sec. 3-4.4, transmit-receive switching is done by means of diodes instead of a relay.

### 3-4.4 Transmit-receive Switching

As shown in Fig. 3-35, transmit-receive switchover is controlled by means of diodes $D_8$, $D_{11}$, and $D_{19}$. During receive, these diodes are not connected to ground and have no effect on the circuit.

During transmit, when the microphone PTT switch is pressed, diodes $D_8$, $D_{11}$, and $D_{19}$ are returned to ground, causing the diodes to conduct. Diode $D_8$ grounds the receiver local oscillator $TR_{11}$ base through $R_{36}$, disabling $TR_{11}$. Diode $D_8$ also grounds receiver AF amplifier $TR_6$ through $R_{109}$ to disable any receiver audio to the amplifier/modulator circuit. Diode $D_{11}$ grounds the RF and IF sections of the receiver. Diode $D_{19}$ switches the level meter over to the power output mode from the S-meter mode. Also, when the PTT switch is pressed, the ground return is completed for the transmitter oscillator $TR_{17}$, transmitter RF amplifier $TR_{19}$, microphone amplifier $TR_{14}$, and the modulation lamp circuit, thus

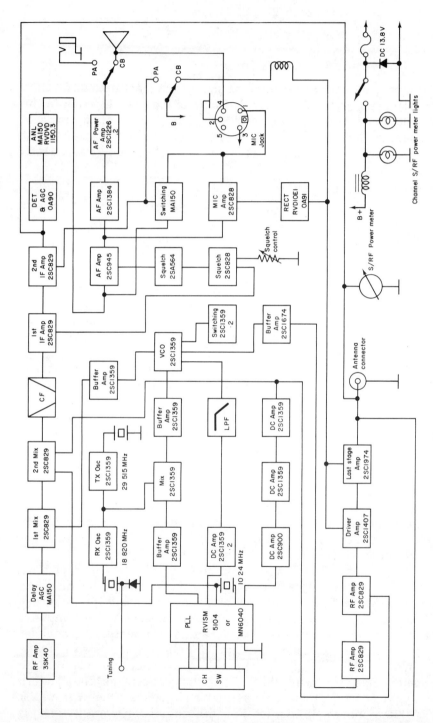

**Figure 3-33:** Phase-locked loop CB block diagram

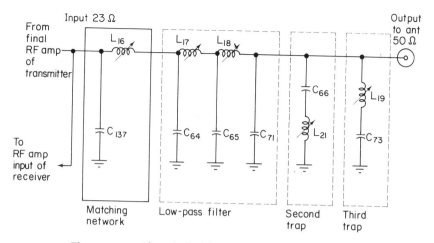

**Figure 3-34:** Phase-locked loop CB RF output filter circuits

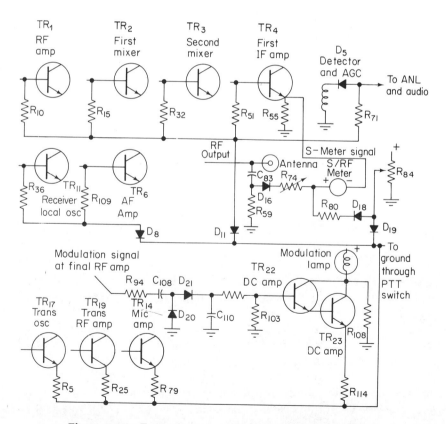

**Figure 3-35:** Transmit-receive switching circuits using diodes

activating all of these stages. The modulation lamp circuit $D_{20}$, $D_{21}$, $TR_{22}$, $TR_{23}$ samples the modulation signal at the collector of the final RF amplifier and drives the modulation lamp accordingly. (Increased modulation produces a brighter glow on the lamp.)

### 3-4.5 Protector and RF Overload Circuits

As shown in Fig. 3-36, the receiver input circuit is protected by diodes. When noise spikes occur (such as lightning, high-level interference, static electricity, etc.), diodes $D_1$ and $D_2$ conduct in order to prevent damage to the RF amplifier $TR_1$. Diode $D_3$ conducts during periods of high noise level or when strong RF signals are received. This reduces the overall output of the RF amplifier to the remaining receiver circuits.

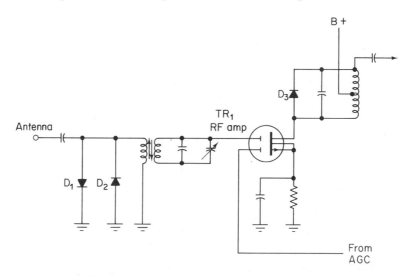

**Figure 3-36:** Protector and RF overload circuits

### 3-4.6 Microphone AGC Circuit

As shown in Fig. 3-37, the set is protected from overmodulation by an AGC circuit consisting of $D_{14}$, $D_{15}$, $C_{78}$, $C_{67}$, $R_{42}$, and $R_{43}$. Audio signals from the microphone are applied through microphone amplifier $TR_{14}$ to the modulation circuit $TR_7$, $TR_8$, $TR_9$, and transformer $T_7$. A sample of the modulation signal is sampled at the junction of $R_{42}$–$R_{43}$ and is rectified by $D_{14}$–$D_{15}$. As the modulation signal increases in amplitude, diodes $D_{14}$–$D_{15}$ conduct more heavily, producing a large negative feedback voltage to the microphone amplifier $TR_{14}$. This negative voltage is reverse bias for $TR_{14}$ (an NPN) and causes the gain of $TR_{14}$ to be reduced. In this manner, modulation is held to a maximum of 100%.

Sec. 3-4    Phase-locked Loop (PLL) CB Circuits                                          165

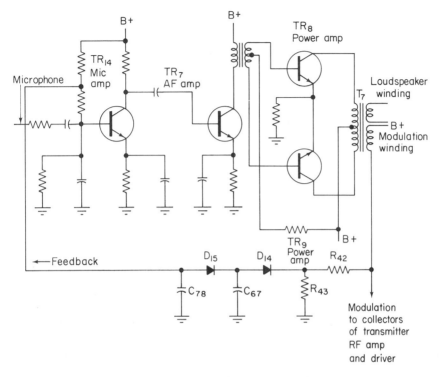

Figure 3-37: Microphone AGC circuits

### 3-4.7 Noise Blanker Circuit

As shown in Fig. 3-38, the noise blanker circuit is somewhat unique in that it uses an FET as the noise amplifier. Under normal operating conditions, the noise blanker has no effect on the receiver. When larger noise spikes are received, they are applied to the gate of noise amplifier $TR_{12}$. The amplified output from $TR_{12}$ is applied to the base of noise blanker $TR_{13}$. This turns on $TR_{13}$ (which is normally biased off), and the collector voltage of $TR_{13}$ drops. The negative output pulse is coupled through $C_{87}$ to the anode of noise blanking diode $D_4$. The negative pulse reverse biases $D_4$, turning off $D_4$ for the duration of the pulse. Since all received signals must pass through $D_4$, the front end of the receiver is effectively muted during this time, and the noise signals or spikes do not pass.

Switch $S_3$ is the noise blanker ON/OFF switch. When the switch is OFF (open), the emitter of $TR_{13}$ has no path to ground and is disabled. Diode $D_9$ is a positive shunt ground, so that only negative-going noise spikes will trigger the noise blanker circuit.

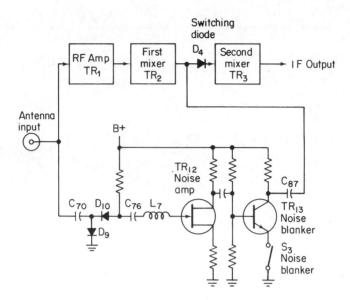

**Figure 3-38:** Noise blanker circuits

### 3-4.8 Automatic Noise Limiter (ANL) Circuit

As shown in Fig. 3-39, the ANL circuit uses diodes instead of transistors. The input to the detector diode $D_5$ is the 455 kHz amplitude-modulated IF carrier. Diode $D_5$ detects the negative envelope of the IF carrier, both audio and noise signals. The detected signal is applied to the ANL diode $D_6$.

Under normal operating conditions, $D_6$ is forward-biased by the voltage divider consisting of $R_{76}$ and $R_{77}$, and the bias is sufficient to allow audio signals to pass. If excessive noise is present with the audio signal, the detected noise voltage drives the anode of $D_6$ negative, and $D_6$ no longer conducts, clipping the audio signal off. The ANL switch shorts out $D_6$ when set to ON (switch closed), eliminating the ANL function.

### 3-4.9 Squelch Circuit

As shown in Fig. 3-40, the squelch circuit is composed of two transistors, $TR_{15}$ and $TR_{16}$. With no signal input, very little negative AGC voltage is developed and applied to the base of $TR_4$ (the first IF amplifier). This means that $TR_4$ conducts heavily, and the emitter of $TR_4$ is at a high positive voltage. The positive voltage is applied to $TR_{15}$ and causes $TR_{15}$ to conduct. With $TR_{15}$ conducting, its collector voltage drops, driving the $TR_{16}$ collector toward $B+$. The high positive voltage is fed to the emitter of $TR_6$ and, since $TR_6$ is NPN, the positive voltage biases $TR_6$ off. Since $TR_6$ is the receiver audio amplifier, no sound can pass to the loudspeaker when $TR_6$ is biased off.

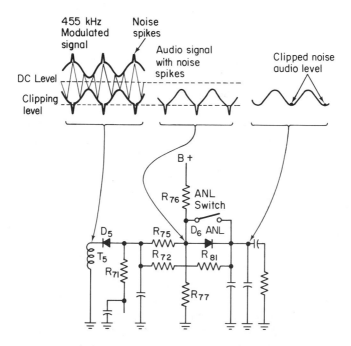

**Figure 3-39:** Automatic noise limiter circuits

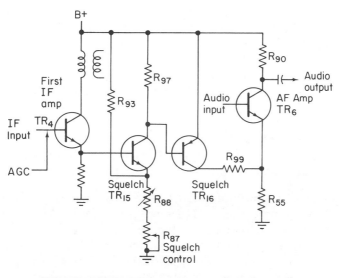

**Figure 3-40:** Squelch circuits

When a strong signal is received, an AGC voltage is developed to cut down the receiver gain. In doing so, $TR_4$ does not conduct as hard, the $TR_4$ emitter is less positive, $TR_{15}$ stops conducting, and the $TR_{16}$ collector is no longer at B+. Under these conditions, the receiver audio amplifier $TR_6$ passes audio signals.

$R_{93}$, $R_{88}$, and $R_{87}$ are the voltage dividers that determine at what point the squelch circuit operates. $R_{87}$ is the front panel SQUELCH control, and $R_{88}$ is an internal servicing adjustment that determines the range of $R_{87}$.

## 3-5 SINGLE SIDEBAND (SSB) CB CIRCUITS

Many present-day CB sets provide for transmission and reception of SSB signals. Generally, this is in addition to conventional AM communications, although there are some SSB-only sets still in existence. One obvious advantage of SSB is that the number of channels is automatically increased. In a set of capable of 23-channel AM operation, the number of channels is increased to 69 if the set includes SSB, since SSB uses the upper sideband (USB) and lower sideband (LSB) of the 23 channels. Thus, you get two extra sideband channels for each AM channel. (On a 40-channel set, you get a total of 120 channels).

Another advantage of SSB is that you are allowed extra power output (12 W PEP for SSB instead of 4 W for AM). Thus, combined with the use of a sideband only (sideband power is essentially lost in AM transmission) you generally get better communications with SSB.

From a technical standpoint, operation of an SSB transceiver is quite different from that of an AM transceiver. For that reason, we shall discuss SSB circuits in full detail. However, as discussed in Chapter 4, the same basic troubleshooting and service techniques used for AM sets may be applied to SSB sets.

### 3-5.1 Basic SSB Theory

Figure 3-41 is the block diagram of the basic SSB transmission system. The microphone signals are amplified and fed to one input of a *balanced modulator* circuit. The other input to the balanced modulator is a fixed-frequency signal from an oscillator. These two signals are mixed, producing upper and lower sidebands. Double sideband (DSB) signals are present at the output of the balanced modulator, but the carrier signal is balanced out.

The balanced modulator is essentially a bridge circuit, and the RF signal from the oscillator is always present. Audio is present only when the microphone is used. When no modulating signal is present, no current flows, since $R_3$ is adjusted to divide or balance the RF input

Sec. 3-5  Single Sideband (SSB) CB Circuits               169

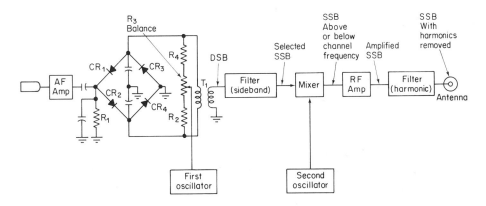

Figure 3-41:  Basic SSB transmission system

across the bridge. Consequently, there is no RF across the primary of $T_1$ when audio is absent.

When audio is present, a positive-going audio signal causes $CR_2$–$CR_4$ to conduct, whereas $CR_1$–$CR_3$ are reverse biased. Negative-going audio produces the opposite results ($CR_1$–$CR_3$ on, $CR_2$–$CR_4$ off). Both positive and negative audio signals thus unbalance the bridge and amplitude modulate the oscillator RF signal to produce upper and lower sidebands, but no carrier.

Both sideband signals are applied through $T_1$ to the filter, which removes one sideband. The lower sideband is removed when USB operation is desired, and vice versa. The SSB signal from the filter is mixed with a fixed-frequency signal from a second oscillator to produce a single sideband signal that is above or below the channel frequency. This SSB signal is fed through the usual RF amplifiers and output harmonic filters to the antenna.

As an example, assume that channel 1 is used, and the audio is in the range of 300 to 3,000 Hz (typical voice frequency range). The channel 1 26.965 carrier will be absent with SSB. When USB is selected, the radiated RF sideband signals are in the range of 26.965300 to 26.968000 MHz. When LSB is selected, the output is in the range of 26.962000 to 26.964700 MHz.

The SSB receiver is essentially the same as an AM receiver, with two major exceptions. First, the carrier removed by the balanced modulator in the transmitter must be restored. Generally this is done by mixing the received signal with a fixed-frequency signal at some point after the IF and before the AM detector.

When the fixed oscillator signal is mixed with the sideband passing through the IF, only the audio remains. The audio, is processed in the same manner as the AM audio signals.

An SSB receiver also includes a *clarifier* control, which is comparable to the delta tune control in AM sets. The clarifier shifts the local oscillator frequency as necessary so that the audio signal will sound natural. By shifting the oscillator frequency, the missing carrier frequency and the frequencies within the sideband are matched to produce a normal sound (rather than something like a voice from outer space).

### 3-5.2  Typical SSB Set

The operation of circuits used in the Pace CB Sidetalk 1000M/1000B sets is discussed below. Both sets are 23-channel, crystal-controlled AM and SSB transceivers. The 1000M is for mobile use, whereas the 1000B is a base station set. The 1000M operates from a 12 V dc power source. The 1000B may be operated from either 117 V ac or 12 V dc power. The 1000B also contains a clock/timer that permits the set to be turned on at a desired time. In addition to measuring S/RF in the usual manner, the 1000B meter also measures SWR. The microphones are low impedance dynamic type; the loudspeakers are 8 Ω. RF power output is 4 W for AM and 12 W PEP for SSB into 50 Ω. Receiver input sensitivity is 0.5 $\mu$V for 10dB [(S+N)/N] with SSB and 1 $\mu$V for 10 dB [(S+N)/N] for AM. Audio output is 3 W, which may be applied to an external loudspeaker for a PA function. The sets include such functions as noise blankers, automatic output limiting, RF gain control, clarifier, and S/RF/SWR meters.

As in AM sets, different combinations of synthesizer frequencies are used to obtain the proper frequencies for AM and SSB operation in both transmit and receive. To simplify the circuit descriptions, a separate block diagram is used for all three transmit functions (AM, USB, LSB). Receiver circuitry is straightforward, and a single block diagram is used for all functions. Whenever necessary, a simplified partial schematic is presented for individual circuit descriptions. Before going into overall descriptions, it is necessary to understand the operation and relationship of the oscillators during the various modes of operation.

### 3-5.3  Oscillators

Three oscillators and 11 crystals are used to produce all of the desired frequencies. As is shown in Fig. 3-42, two of the oscillators and 10 crystals are used in the synthesizer. The synthesizer also receives a 15.605 MHz signal from the oscillator/doubler circuit shown in Fig. 3-43. The oscillator/doubler $Q_{23}$ also provides 7.8025 MHz signals to the balanced modulator and product detector (which is the detector used in

Sec. 3-5 Single Sideband (SSB) CB Circuits    171

SSB reception). The 11.8 MHz oscillator $Q_{11}$ is controlled by one of six crystals; the 7.4 MHz oscillator $Q_{10}$ is controlled by one of four crystals.

The combined outputs of $Q_{10}$ and $Q_{11}$ are mixed in 19 MHz mixer $Q_9$. The output of $Q_9$ is applied to one gate of FET mixer $Q_7$ and to the input of 19 MHz amplifier $Q_8$. The other gate of $Q_7$ receives a 15.605 MHz signal from $Q_{23}$ (Fig. 3-43). The output of $Q_7$ is amplified by $Q_6$ and mixed with the output of $Q_8$.

All of the oscillators are crystal controlled, and each crystal is provided with a separate trimmer adjustment. This makes it possible to set each crystal to a precise frequency during service. The actual frequency of the oscillators is determined by the crystal when the channel selector and AM/USB/LSB switch is set to the desired position. However, the exact frequencies may be changed slightly in one of two ways.

The $Q_{11}$ oscillator frequency may be shifted by the front panel clarifier control $R_{229}$. As discussed in Sec. 3-5.1, this is necessary to produce a normal voice sound during reception. Note that $R_{229}$ is not used in the circuit during transmit. Instead, internal adjustment $R_{228}$ replaces front-panel $R_{229}$. The full range of the clarifier is about $\pm 500$ Hz.

The $Q_{10}$ oscillator frequency is shifted (increased) by 2,500 Hz during AM receive, in order to make the AM signals compatible with the 7.8025 MHz signals from $Q_{23}$ (Fig. 3-43) and a 7.8 MHz filter. Diode $CR_{204}$ (Fig. 3-42) is forward-biased in all transmit modes and SSB receive. This shunts out the network $C_{225}$, $C_{226}$, $C_{227}$, $L_{204}$, and $R_{223}$. Under these conditions, $Q_{10}$ produces a signal at the selected crystal frequency. In AM receive, $CR_{204}$ is reverse-biased by removal of voltages. (These voltages are applied and removed as necessary when the AM/USB/LSB and PTT switches are operated.) With $CR_{204}$ reverse-biased, the network is connected to crystals, and the $Q_{10}$ oscillator frequency is increased by 2,500 Hz.

The oscillator frequencies for one channel (channel 23) are shown in Fig. 3-44. The oscillator/doubler $Q_{23}$ frequency remains at 7.8025 MHz for all channels and all modes of operation. The remaining oscillator frequencies are selected as necessary, depending upon the channel and mode of operation. For example, as discussed in Sec. 3-5.6, when AM receive is used on channel 23, the 19.455 MHz synthesizer output is mixed with the incoming 27.255 MHz signal to produce an IF of 7.8 MHz. As discussed in Sec. 3-5.4, when AM transmit is used on channel 23, the 19.4525 MHz synthesizer output is mixed with the 7.8025 carrier signal to produce an output 27.255 MHz. The same combinations are used for LSB transmit and receive, as discussed in Sec. 3-5.5. However, for USB transmit and receive, the 35.0575 MHz synthesizer output is mixed with the 7.8025 carrier signal to produce the 27.255 MHz.

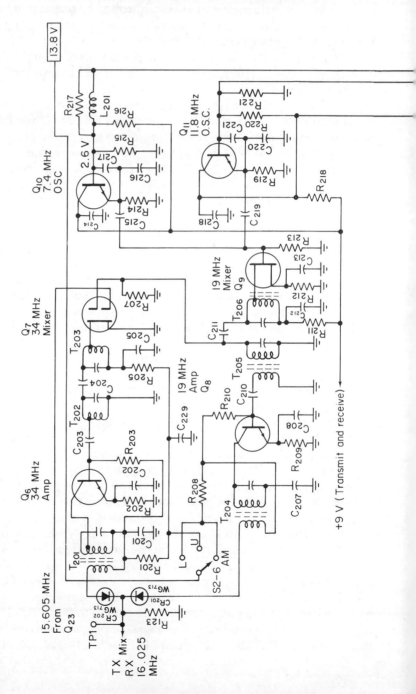

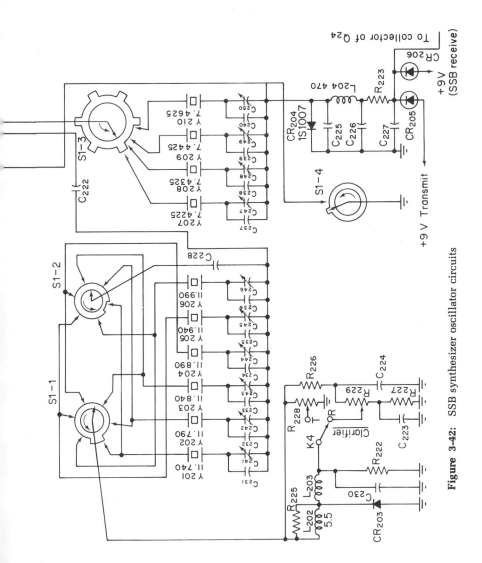

**Figure 3-42:** SSB synthesizer oscillator circuits

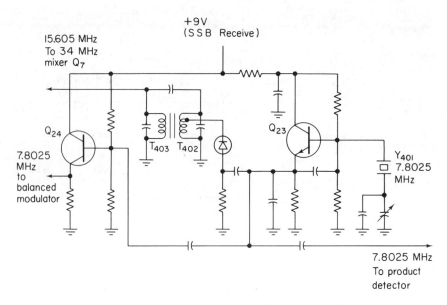

**Figure 3-43:** SSB 7.8025 MHz oscillator and doubler

| Channel | $Q_{10}$ Osc | $Q_{11}$ Osc | 19 MHz Synth (A) | 34 MHz Synth (B) | $Q_{10}$ Osc | $Q_{11}$ Osc | 19 MHz Synth (C) |
|---------|--------------|--------------|------------------|------------------|--------------|--------------|------------------|
| 23 | 7.4625 | 11.990 | 19.4525 | 35.0575 | 7.465 | 11.990 | 19.455 |

All frequencies in MHz
(A) Used in LSB transmit, LSB receive, and AM transmit
(B) Used in USB transmit and receive = (LSB receive + (2) 7.8025)
(C) Used in AM receive = (2.5 kHz + AM transmit)

**Figure 3-44:** SSB synthesizer frequencies for channel 23

Sec. 3-5  Single Sideband (SSB) CB Circuits                                      175

### 3-5.4  AM Transmitter

Figure 3-45 is the block diagram of the circuits used during AM transmit. As shown, signals from the 19 MHz synthesizer are fed to one gate of FET mixer $Q_5$. The other gate of $Q_5$ receives a 7.8 MHz signal from $Q_{23}$. The 27 MHz output from $Q_5$ is amplified by bandpass amplifier $Q_4$ and predriver $Q_3$, and is fed to driver $Q_2$ and power amplifier $Q_1$, where the signal is modulated and coupled to the antenna.

During AM operation, the balanced modulator is unbalanced by application of a voltage to one side of the modulator bridge. This permits the 7.8025 MHz carrier signal to pass from $Q_{23}$ and $Q_{24}$ through the modulator, filter $FL_{301}$, and amplifier $Q_{17}$ to $Q_5$. No audio is applied to the balanced modulator during AM transmit. Thus, the 7.8025 MHz signal is CW (continuous wave).

*Modulator.* Audio signals from the microphone are coupled to $Q_{25}$, which functions as an automatic gain control during AM transmission. The audio is then amplified by $Q_{26}$ and $Q_{30}$, which function as an active low-pass filter, and is further amplified by $IC_{501}$, which is an IC amplifier/modulator. The audio from $IC_{501}$ directly modulates $Q_2$ and $Q_1$.

*Modulation limiter.* Part of the audio modulating signals from $IC_{501}$ are fed back through $R_{616}$ and rectified by $CR_{603}$. The rectified dc is then filtered and fed to the gate of $Q_{25}$ as an adjustable bias. This bias is adjusted by $R_{616}$ to set $Q_{25}$ gain (and thus the gain of the entire audio modulation section), so that modulation does not exceed 100%.

### 3-5.5  SSB Transmitter

Figures 3-46 and 3-47 are the block diagrams for the LSB and USB transmitter circuits, respectively. Note that the circuits are essentially the same except for the oscillator and synthesizer arrangement.

*LSB transmitter.* As shown in Fig. 3-46, audio signals from the microphone are amplified in $Q_{25}$ and $Q_{26}$ and coupled to the balanced modulator. This audio signal modulates the 7.8025 carrier frequency to produce a suppressed carrier double sideband (DSB) signal, as discussed in Sec. 3-5.1. One of the two sidebands is cut off by crystal filter $FL_{301}$, and the other sideband appears as a suppressed carrier SSB signal.

The SSB signal is amplified by $Q_{17}$ and coupled to one gate of transmitter mixer $Q_5$. The other gate of $Q_5$ receives the 19 MHz synthesized signal from $Q_8$ and $Q_9$. (The actual frequency combinations for channel 23 are given in Fig. 3-44.) An LSB-modulated 27 MHz signal

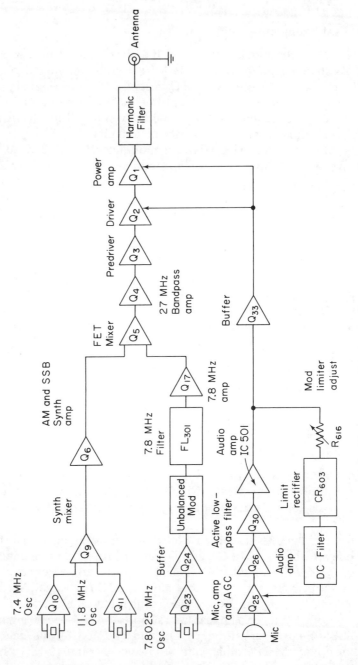

**Figure 3-45:** AM transmitter block diagram

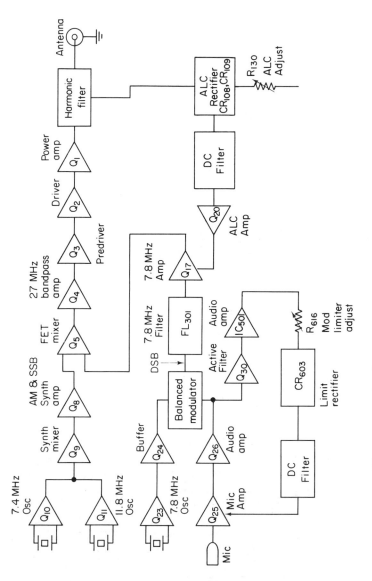

**Figure 3-46:** SSB lower sideband (LSB) transmitter block diagram

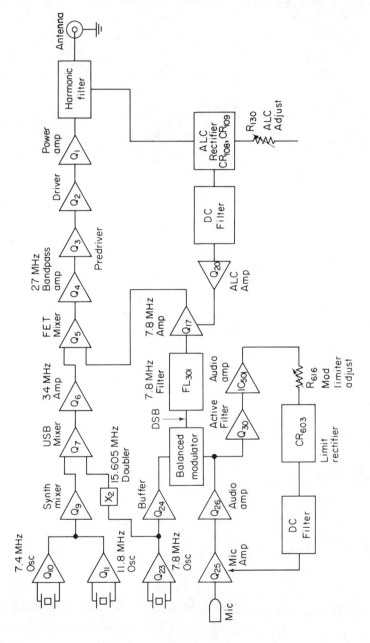

**Figure 3-47:** SSB upper sideband (USB) transmitter block diagram

Sec. 3-5 Single Sideband (SSB) CB Circuits 179

appears at the output of $Q_5$, where the signal is amplified by $Q_4$ and $Q_3$. The modulated output is then fed to the antenna through $Q_2$, $Q_1$, and the harmonic filter. Note that $Q_1$ through $Q_4$ are wideband linear amplifiers capable of passing the modulated RF signals.

*Automatic level control (ALC).* During SSB operation, RF power output is limited to 12 W PEP by the ALC circuits. Part of the RF power output at the harmonic filter is fed back and rectified by $CR_{108}$–$CR_{109}$. The rectified dc is then filtered and fed to the base of $Q_{17}$ through $Q_{20}$ as an adjustable bias. ($Q_{20}$ operates as an AGC circuit during receive.) The bias on $Q_{17}$ is adjusted by the ACL adjust control $R_{130}$ to set $Q_{17}$ gain (and thus the amplitude of the RF output) so that output does not exceed 12 W PEP.

*USB transmitter.* As shown in Fig. 3-47, operation of the USB circuits is essentially the same as that of LSB. However, to obtain USB transmit frequencies, 19 MHz synthesized frequencies are mixed with the *doubled* frequency from the carrier oscillator in $Q_7$ (as discussed in Sec. 3-5.3 and shown in Fig. 3-43). This produces a 34 MHz synthesized frequency which is mixed in $Q_5$ with the suppressed carrier SSB from $Q_{17}$. The difference frequency produces the upper sideband, which is then amplified and fed to the antenna (as described for LSB).

### 3-5.6 Receiver

Figure 3-48 is the block diagram of the circuits used during AM, LSB, and USB receive. Except for the synthesizer frequencies used, the only major difference between AM and SSB during receive is the detector system. A conventional AM detector and noise limiter is used for AM receive. A product detector $Q_{21}$ is used for SSB receive to mix the reintroduced carrier from $Q_{23}$ with the USB or LSB output from the IF amplifiers.

*RF amplifier and mixer.* RF signals from the antenna pass through an antenna switching circuit (Sec. 3-5.7) and are coupled to RF amplifier $Q_{12}$. After amplification, the RF signals are fed to one gate of RF mixer $Q_{13}$, where they are mixed with local oscillator signals from the synthesizer.

*Local oscillator frequencies.* Oscillator circuits are discussed in Sec. 3-5.3, and synthesizer frequencies for channel 23 are shown in Fig. 3-43. Note that the clarifier that shifts the local oscillator frequencies is used only during SSB.

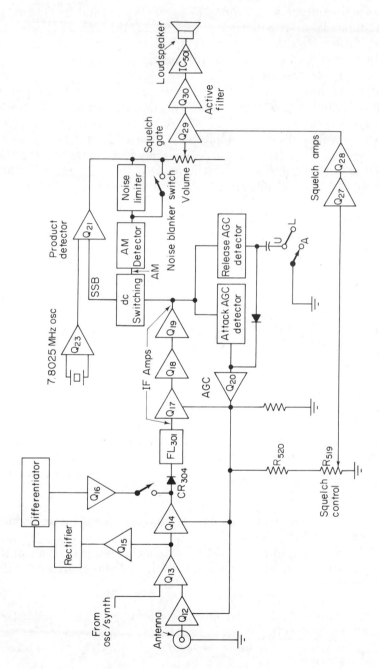

**Figure 3-48:** AM and SSB receiver block diagram

Sec. 3-5  Single Sideband (SSB) CB Circuits                                181

*IF amplifier.* IF signals from the mixer are fed into the crystal filter $FL_{301}$, which passes only the desired signals and cuts out the undesired signals. In AM receive, the carrier plus one sideband are passed through $FL_{301}$. In SSB, only the sideband passes $FL_{301}$.

The output of $FL_{301}$ is amplified by $Q_{17}$, $Q_{18}$, and $Q_{19}$, and is then applied to the detector. For AM receive, the IF output is applied through a switching network to the AM detector and noise limiter (Fig. 3-49). For SSB receive, the IF is mixed with the 7.8025 MHz signal from $Q_{23}$ in product detector $Q_{21}$ (Fig. 3-50).

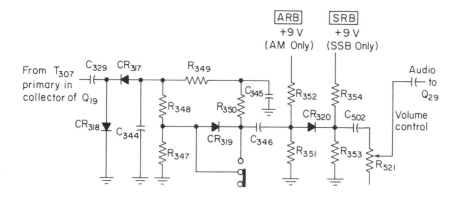

**Figure 3-49:** AM detector and noise limiter circuits

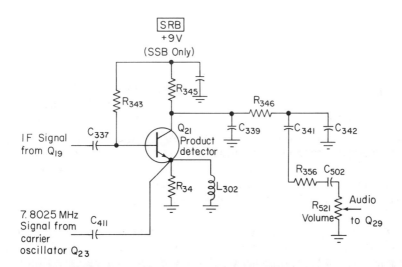

**Figure 3-50:** SSB detector circuits

*AM detector and noise limiter.* As shown in Fig. 3-49, the IF signal from $Q_{19}$ is fed to envelope detector diodes $CR_{317}$ and and $CR_{318}$. The detected audio is applied through the noise limiter and the volume control $R_{521}$ to $Q_{29}$.

Noise limiting is achieved with a network consisting of $R_{347}$ through $R_{350}$, $C_{345}$, and $CR_{319}$. Direct current bias from the detector is applied to the anode of $CR_{319}$ at the junction of divider $R_{347}$–$R_{348}$. The dc bias is also applied to the cathode of $CR_{319}$ via $R_{349}$ and $R_{350}$. This forward-biases $CR_{319}$ for normal signal amplitudes, and the audio is coupled through $CR_{319}$ and $R_{521}$ to $Q_{29}$.

When noise pulses are present in the signal, a higher negative bias is applied to the anode of $CR_{319}$. However, bias to the cathode is not increased because of the time constant presented by $R_{349}$ and $C_{345}$. This reverse-biases $CR_{319}$ so that the noise pulses are clipped off.

Diode $CR_{320}$ is a dc keying switch. In AM receive mode, $CR_{320}$ is forward-biased by the ARB voltage applied to the anode. Audio from the AM detector and limiter then passes through $R_{521}$ to $Q_{29}$. When SSB receive is selected, $CR_{320}$ is reverse-biased by the SRB voltage applied to the cathode, and audio from the AM detector and limiter is blocked.

*SSB detector.* As shown in Fig. 3-50, the IF signal from $Q_{19}$ is applied to the base of $Q_{21}$. The emitter of $Q_{21}$ receives a 7.8025 MHz signal from the carrier oscillator. Detected audio from the collector of $Q_{21}$ is fed through volume control $R_{521}$ to $Q_{29}$. Product detector $Q_{21}$ is active only in SSB mode when the SRB voltage is applied.

*Noise blanker.* As shown in Fig. 3-51, the noise blanker consists of a rectifier, a differentiator, a noise amplifier, and a gate diode.

Impulse noise is picked up at the antenna and amplified by $Q_{12}$. The noise is converted to a 7.8 MHz IF by the mixer $Q_{13}$ and further amplified by $Q_{15}$. Positive peaks of the impulse noise are rectified by $CR_{307}$–$CR_{308}$ and differentiated by $C_{317}$ and $R_{319}$. The differentiated pulses are then coupled to the gate of noise amplifier $Q_{16}$, where they are amplified to a sufficient level to control gate diode $CR_{304}$. When strong impulse noise is present in the received signal, $CR_{304}$ becomes reverse-biased and prevents the noise from passing through to the IF. When the noise blanker switch is in the OFF position, bias is removed from $Q_{16}$, and the circuit has no control of $CR_{304}$ (all signals pass).

*Automatic gain control (AGC).* Since there is no carrier present on the SSB signal, special circuitry is required to maintain the AGC voltage at a proper level for both very weak and very strong signals. As shown in Fig. 3-52, two AGC detector circuits are used.

One circuit, for weak signals, is called an *attack* AGC. IF signals from

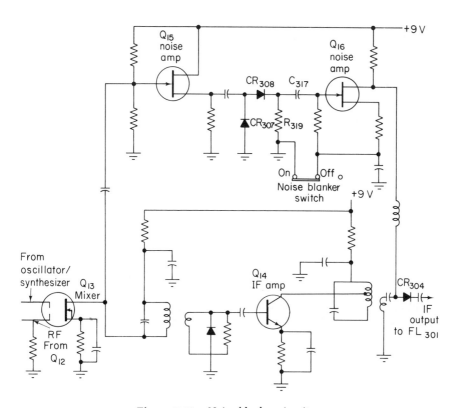

**Figure 3-51:** Noise blanker circuits

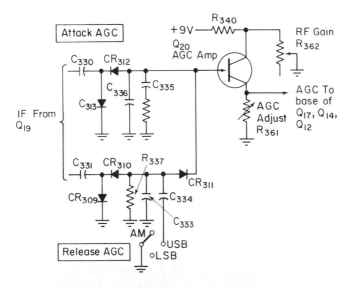

**Figure 3-52:** Automatic gain control (AGC) circuits

$Q_{19}$ are applied to the attack AGC circuit through $C_{330}$. These signals are rectified by $CR_{312}$–$CR_{313}$ and applied to AGC amplifier $Q_{20}$. The other circuit is called the *release* AGC. IF signals from $Q_{19}$ are applied to the release AGC circuit through $C_{331}$. These signals are rectified by $CR_{309}$–$CR_{310}$ and applied to $Q_{20}$ through $CR_{311}$. The AGC feedback voltage to the bases of $Q_{12}$, $Q_{14}$, and $Q_{17}$ is taken from the source of $Q_{20}$. The level of the AGC voltage is set by $R_{361}$ and $R_{362}$. During adjustment, $R_{362}$ is set at maximum and $R_{361}$ is adjusted to produce the desired AGC voltage. Then $R_{362}$, which is the front-panel RF GAIN control, is used to vary the AGC bias and thus set the desired amount of RF gain.

When a strong signal is received, attack diodes $CR_{312}$ and $CR_{313}$ work quickly and charge $C_{335}$–$C_{336}$. At the same time, release AGC diodes $CR_{309}$ and $CR_{310}$ charge $C_{334}$. Note that $C_{334}$ is returned to ground only when USB or LSB receive is selected. Thus, $C_{334}$ has no effect on the circuit during AM receive. The charge on $C_{334}$ reverse-biases $CR_{311}$, which remains in this condition until $C_{334}$ has fully discharged through $R_{337}$. This delay period is necessary to allow sufficient time for $C_{335}$–$C_{336}$ to discharge, thus maintaining a constant AGC bias in the presence of very strong or very weak signals (within limits).

*Squelch.* As shown in Fig. 3-53, squelch control $R_{519}$ sets the operating bias of squelch amplifiers $Q_{27}$ and $Q_{28}$. Squelch gate $Q_{29}$ controls the flow of audio from the detector (AM or SSB) to the audio amplifier.

When no signal is being received, $R_{519}$ is set to forward-bias $Q_{27}$, which turns on $Q_{28}$. This places a positive voltage on the emitter of $Q_{29}$, which gates off the audio. When a strong signal is received, the negative

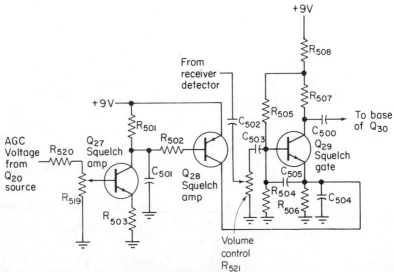

**Figure 3-53:** Squelch circuits

Sec. 3-5  Single Sideband (SSB) CB Circuits                    185

AGC voltage overrides the voltage set on $Q_{27}$ by $R_{519}$, cutting off $Q_{27}$. In turn, this cuts off $Q_{28}$, removing the positive voltage from the emitter of $Q_{29}$, and the audio is gated on. Thus, the voltage set by $R_{519}$ determines the squelch point of the receiver. $R_{520}$ adjusts the range of $R_{519}$.

*Audio.* Returning to Fig. 3-48, the audio signal from squelch gate $Q_{29}$ is coupled to the audio amplifier through active low-pass filter $Q_{30}$. Intergrated circuit $IC_{501}$ is a self-contained audio amplifier consisting of a driver and push-pull final stage. The output of $IC_{501}$ is coupled through a matching transformer to the speaker.

### 3-5.7 Antenna Switching

The antenna switching is accomplished by dc switching diodes instead of relay contacts, and is thus "all electronic switching." This electronic switching eliminates the possibility of RF leakage through the switch.

In the normal receive mode, as shown in Fig. 3-54, +9 V is applied to

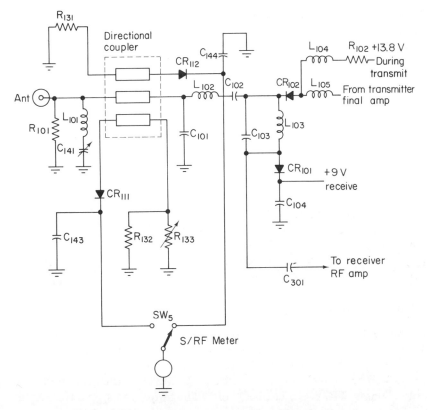

**Figure 3-54:** Antenna switching and S/RF meter circuits

the cathode of $CR_{101}$, thus reverse-biasing $CR_{101}$. At the same time, +13.8 V is removed from the anode of $CR_{102}$, reverse-biasing $CR_{102}$. With $CR_{102}$ reverse-biased, the antenna is effectively disconnected from the transmitter final output stage. Simultaneously, the receiver RF input stage is connected to the antenna via $C_{301}$, $C_{103}$, $C_{102}$, and $L_{102}$.

In transmit, when the PTT switch is pressed, +13.8 V is applied to the anode of $CR_{102}$, forcing $CR_{102}$ into conduction. The path is then open from the transmitter final RF output stage to the antenna via $CR_{102}$, $C_{102}$, and $L_{102}$. Simultaneously, +9 V is removed from the cathode of $CR_{101}$, forward-biasing $CR_{101}$. With $CR_{101}$ conducting, RF input to the receiver is shunted to ground through $C_{301}$, $CR_{101}$, and $C_{104}$, effectively disconnecting the antenna from the receiver input.

Note that the antenna is connected to the antenna switching circuits through a directional coupler, such as that described in Sec. 2-11 of Chapter 2. This arrangement applies only to the 1000B set and permits the S/RF meter to be used as an SWR meter.

### 3-6  WALKIE-TALKIE CB CIRCUITS

Figures 3-55 and 3-56 are the block and schematic diagrams, respectively, for the Pace CB155 hand-held (walkie-talkie) CB set. The following paragraphs describe operation of the circuits.

#### 3-6.1  General Description

The CB155 is a six-channel, crystal-controlled AM transceiver. The set is all solid-state and is operated on a 12 V dc internal rechargeable battery. The RF power input is switchable to either 1 W or 5 W. A telescoping rod antenna is built in, and an external antenna jack is provided. A single device serves as loudspeaker (during receive) and microphone (during transmit). An external microphone and loudspeaker can be connected to the audio section, thus providing a public address (PA) function. Modulation is 85 to 100%. Receiver input sensitivity is 0.5 $\mu$V for 10dB [(S+N)/N]. The set includes such functions as squelch, volume, and S/RF meter.

#### 3-6.2  Receiver

The CB155 receiver is a single-conversion superheterodyne type, using a crystal-controlled modified Colpitts oscillator. An RF signal from the antenna is fed through $L_2$ to $Q_1$. (Note that the transistors shown in Fig. 3-56 are listed as TR, rather than Q. However, the same numbers apply. That is, $TR_1$ is $Q_1$, $TR_2$ is $Q_2$, etc.) The amplified signal from $Q_1$ is

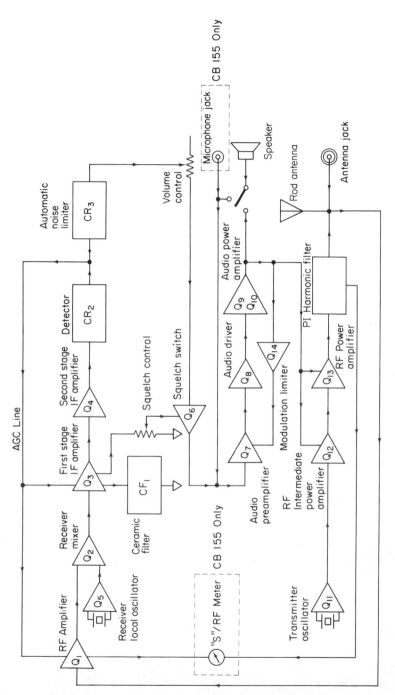

**Figure 3-55:** Pace CB155 walkie-talkie CB block diagram

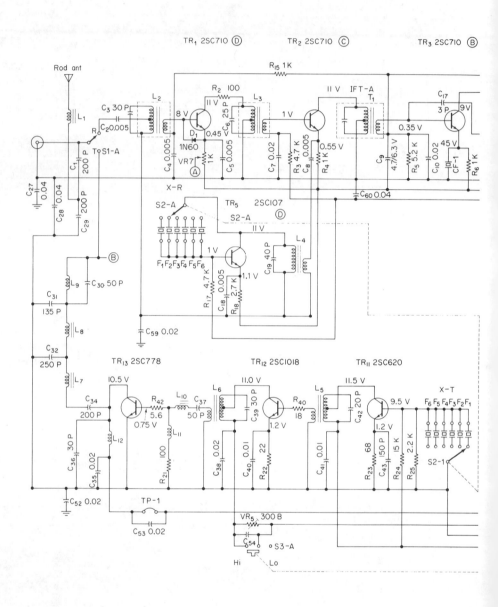

**Figure 3-56:** Pace CB155 walkie-talkie CB schematic diagram

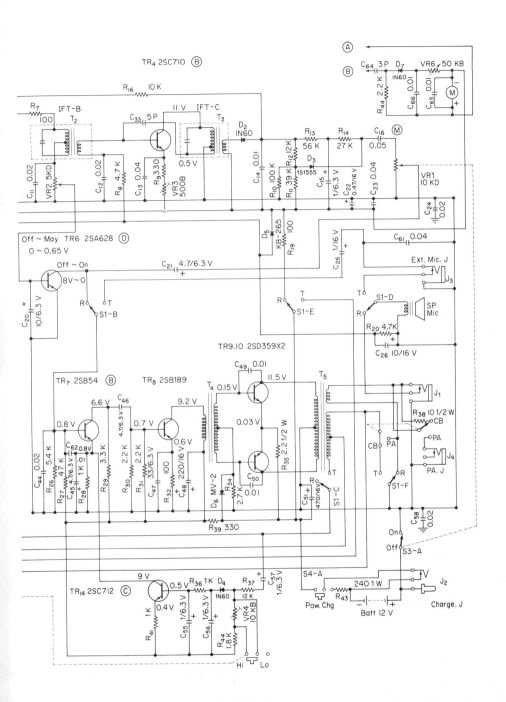

mixed at $Q_2$ with the local oscillator output from $Q_5$. The output of $Q_2$ is at the IF frequency of 455 kHz and is coupled to the first IF amplifier $Q_3$ through $T_1$.

At $Q_3$, the signal is amplified and filtered by ceramic filter $CF_1$, located in the emitter of $Q_3$. Filter $CF_1$ improves the selectivity (which is about $-41$ dB at 10 kHz), and image rejection (which is 20 dB). The filtered signal is fed to $Q_4$ through $T_2$, where the IF signal is again amplified. The IF signal is coupled through $T_3$ to the detector $CR_2$, which removes the audio from the IF signal.

The audio signal is passed through automatic noise limiter diode $CR_3$ (shown as $D_3$ on Fig. 3-56) to volume control $VR_1$. The audio signal is controlled by the squelch switch $Q_6$. When switch $Q_6$ is off, the signal is amplified at preamplifier $Q_7$ and again at $Q_8$, which supplies enough power to drive the push-pull audio amplifier $Q_9$ and $Q_{10}$.

When a signal above the desired squelch point is present at $Q_3$, the squelch switch $Q_6$ is biased off, depending upon the setting of the squelch control $VR_2$. With $Q_6$ off, the audio signal is not shorted to ground. When there is no signal, or the signal is below the desired squelch point, $Q_6$ is biased on and pulls the audio input to $Q_7$ to ground, thus squelching the audio.

### 3-6.3 Transmitter

The CB155 transmitter is a three-stage type, using a crystal-controlled modified Colpitts oscillator. Transistor $Q_{11}$ forms the fundamental frequency oscillator, which generates the desired channel frequency. $Q_{12}$ and $Q_{13}$ are the transmitter driver and final stages, respectively. Both RF stages are modulated by transformer $T_5$ (which serves the usual dual purpose of modulation transformer and audio output transformer). The HI/LO switch in the collector circuit of $Q_{12}$ allows the transmitter input power to be switchable from 5 W to 1 W.

Transistor $Q_{14}$ functions as a modulation limiter. A portion of the audio at the output of $T_5$ is fed back through $Q_{14}$ to $Q_7$. Increases in modulation decrease the gain of $Q_7$, thus maintaining the modulation level constant.

The low-pass pi-filter $C_{31}$, $C_{32}$, $L_7$, $L_8$, $L_9$ removes harmonics in the usual manner.

### 3-6.4 S/RF Meter

In the receive mode, the incoming signal forces $Q_1$ towards cutoff. The resultant voltage change across $VR_7$ causes the meter to move up-scale (from $S_1$ to $S_9$). Potentiometer $VR_7$ permits adjustment of the S-meter during receive.

In the transmit mode, a portion of the transmitted signal is tapped off through $C_{64}$ and is rectified and filtered by $CR_7$ ($D_7$) and $C_{66}$. An increase in power output causes the meter to move up-scale. Potentiometer $VR_6$ permits adjustment of the RF meter during transmit.

# 4
# BASIC CB SERVICE APPROACH

In this chapter, we shall describe the basic approach to CB service; that is, how the basic troubleshooting procedures of Chapter 1 are combined with the practical use of test equipment discussed in Chapter 2, together with a knowledge of typical CB circuits covered in Chapter 3, in order to locate specific faults in various types of CB sets.

Since alignment and adjustment, as well as testing, are part of service, we will also describe such basic procedures. Specific alignment, adjustment, and test procedures for typical CB sets are covered in Chapter 6. Throughout this chapter, considerable emphasis is placed on "universal" test, service, and troubleshooting procedures. These procedures apply to all CB sets now in existence, as well as those that may be introduced in the future. This discussion is followed by specific examples of troubleshooting CB set circuits.

In the specific troubleshooting examples, you will note that we describe both the steps that *could be* taken and the steps that *should be* taken after each measurement or observation. Sometimes, the could-be steps are almost as logical as the should-be steps, a common occurrence in troubleshooting CB equipment. A knowledge of the *difference* between the could and should steps generally makes the difference between someone who just knows electronics and someone who is a really good CB service technician.

## 4-1 SERVICING NOTES

The following notes summarize practical suggestions for troubleshooting all types of CB sets.

### 4-1.1 Solid-state Servicing Techniques

Although the techniques described below apply primarily to solid-state CB, they are also generally valid for vacuum-tube sets.

*Transient voltages.* Be sure that power to the set is turned off when making repairs. Of course, you must have power on to measure voltages, trace signals, etc. However, be careful when changing components or inserting new components with the power on. Transistors (and possibly diodes) can be damaged by the transient voltages that may develop.

*Disconnected parts.* When working with any CB, do not operate the transmitter with the antenna disconnected, unless a dummy load is used. Likewise, do not operate with the loudspeaker disconnected. This is particularly important for solid-state CB. If you must disconnect the loudspeaker, replace it with a noninductive resistance of the same value as the speaker impedance. Typically, loudspeaker impedance is 3.2 or 8 $\Omega$, but always consult the schematic and/or parts lists for correct speaker impedance. If you are working with FET transistors, be particularly careful when disconnecting them from the set. In circuit, FETs are as rugged as any similar transistor or diode. Out of circuit, FETs may be damaged by static discharge. For this reason, FETs are generally shipped with a clip or ring that shorts all of the terminals or leads together, thus eliminating discharge between the leads. Leave this ring in place until the FET is reconnected in the circuit. If this is not practical, keep one hand on the chassis when disconnecting and reconnecting FETs.

*Sparks and voltage arcs.* Avoid sparks and arcs when troubleshooting any type of CB, particularly solid-state sets. The transient voltages developed can damage small-signal transistors. For example, some technicians will short directly from the base of a transistor to ground (or what they think is ground) in order to test operation of a transistor stage. This could produce a voltage arc that might damage the transistor. As discussed in other sections of this chapter, if you must test transistors in-circuit, always use a capacitor or resistor, never a direct short.

*Intermittent conditions.* If you run into an intermittent condition and can find no fault using routine checks, try tapping (not pounding) the components (vacuum tubes, transistors, diodes, and so forth). If this does not work for solid-state components, try rapid heating and cooling. A small portable hair dryer and a spray-type circuit cooler make good heating and cooling tools, respectively.

First, apply heat; then, cool the component. The quick change in temperature will normally cause an intermittently defective component to go bad permanently. In many cases, the component will open or short, making it easy to locate.

As an alternate procedure, measure the gain of a transistor with an in-circuit transistor tester (Sec. 2-9). Then, subject the transistor to rapid changes in temperature. If the suspected transistor changes its gain *drastically*, or if there is *no change* in gain, the transistor is probably defective.

In any case, do not hold a heated soldering tool directly on or very near a transistor or diode case; this would probably destroy the transistor or diode.

If time permits, you can locate an intermittently defective transistor by measuring in-circuit gain when the set is cold. Then let the set operate until the trouble occurs, and measure the gain of transistors while they are hot. Some variation will be noted in all transistors, but a leaky transistor has a much lower gain reading when it is hot.

*Operating control setting.* If any transistor or vacuum-tube element appears to have a short (particularly the base or grid), check the settings of any operating controls or adjustment controls associated with the circuits. For example, in the audio/modulation section, the volume control set to zero or minimum can give the same indication as a short from element to ground. This condition is shown in Fig. 4-1.

*Recorded gain readings.* If you service any particular make or model of CB set regularly, record the transistor gain readings of a set that is working properly on the schematic for future reference. Compare these gain readings with the minimum values listed in the service literature.

*Shunting capacitors.* It is common practice in troubleshooting vacuum-tube circuits to shunt a suspected capacitor with a known-good capacitor. This technique is good only if the suspected capacitor is open. The test is of little value if the capacitor is leaking or shorted. In any event, avoid the shunting technique when troubleshooting solid-state circuits. This is especially true with an electrolytic capacitor (often used as an emitter bypass in solid-state CB); the transient voltage surges can

Sec. 4-1   Servicing Notes                                                                 **195**

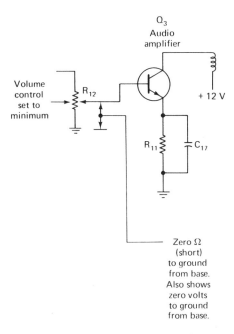

**Figure 4-1:** Example of how operating control setting can affect voltage and resistance readings at transistor elements during troubleshooting

damage transistors. In general, avoid any short-circuit tests with solid-state sets.

*Test connections.* The cases of most metal-case transistors are tied to the collector, so that you can use the case as a test point. Avoid using a clip-type probe on transistors. Also avoid clipping onto some of the subminiature resistors used in solid-state sets. Any subminiature component may break if handled roughly.

*Injecting signals.* When injecting a signal into a circuit (base of a transistor, grid of vacuum tube, input of an IC), make sure that there is a blocking capacitor in the signal-generator output. Most signal generators have some form of blocking capacitor to isolate the output circuit from the dc voltages that may appear in the circuit. In the case of a solid-state CB, the blocking capacitor also prevents the base from being connected to ground (through the generator's output circuit) or from being connected to a large dc voltage (in the generator circuit). Either of these conditions may destroy the transistor. If the generator does not have a built-in blocking capacitor, connect a capacitor between the generator's output lead and the transistor base (or other signal injection point).

### 4-1.2 Measuring Voltages in Circuit

As discussed in Chapter 1, it is possible to locate many defects in CB circuits by measuring and analyzing voltages at the elements of active devices (cathode, grid, plate of vacuum tubes; emitter, base, and collector of transistors). This may be done with the circuit operating and without disconnecting parts. Once located, the defective part can be disconnected and tested or substituted, whichever procedure is most convenient.

Vacuum-tube circuits may be analyzed with a simple VOM or digital meter, whichever you prefer. The normal relationships of vacuum-tube elements are generally fixed. For example, the plate is positive; the cathode is at ground or positive; the grid is (usually) negative.

Transistor circuits are best analyzed with a digital meter or with a very sensitive VOM. A number of manufacturers produce VOMs designed specifically for transistor troubleshooting (the Simpson Model 250 is a typical example). These VOMs have very low voltage scales to measure the differences that often exist between elements of a transistor (especially the small voltage difference between the emitter and base). Such VOMs also have a very low voltage drop (about 50 mV) in the current ranges.

*Analyzing transistor voltages.* Figure 4-2 shows the basic connections for both PNP and NPN transistor circuits; the coupling or bypass capacitors are omitted to simplify the explanation. The purpose of Fig. 4-2 is to establish *normal* transistor-voltage relationships. With a normal pattern established, it is relatively simple to find an abnormal condition.

In practically all transistor circuits, the emitter-base junction is forward-biased to get current flow through the transistor. In a PNP transistor, this means that the base must be made more negative (or less positive) than the emitter. Under these conditions, the emitter-base junction will draw current and cause heavy electron flow from the collector to the emitter. In an NPN transistor, the base must be made more positive (or less negative) than the emitter in order for current to flow from emitter to collector.

The following general rules are helpful when analyzing transistor voltages as part of troubleshooting:

1. The middle letter in PNP and NPN always applies to the base.
2. The first two letters in PNP and NPN refer to the *relative* bias polarities of

Sec. 4-1 Servicing Notes 197

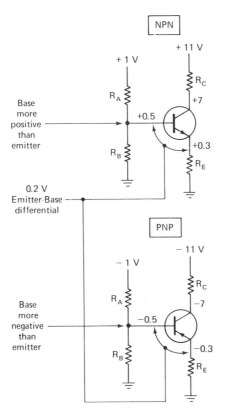

**Figure 4-2:** Basic connections for NPN and PNP transistor circuits (with normal voltage relationships)

the *emitter* with respect to the base or collector. For example, the letters PN (in PNP) indicate that the emitter is positive with respect to both the base and the collector, and the letters NP (NPN) indicate that the emitter is negative with respect to both the base and collector.

3. The collector-base junction is always reverse-biased.

4. The emitter-base junction is usually forward-biased, or is zero-biased in the case of class C amplifiers used in RF circuits.

5. A *base-input* voltage that opposes or decreases the forward bias also decreases the emitter and collector currents.

6. A *base-input* voltage that aids or increases the forward bias also increases the emitter and collector currents.

7. The dc electron flow is always against the direction of the arrow on the emitter.

8. If electron flow is into the emitter, electron flow is out from the collector.

9. If electron flow is out from the emitter, electron flow is into the collector.

Using these basic rules, normal transistor voltages may be summed up this way:

1. For an NPN transistor, the base is positive, the emitter is not quite so positive, and the collector is far more positive.
2. For a PNP transistor, the base is negative, the emitter is not quite so negative, and the collector is far more negative.

*Measurement of transistor voltages.* Discussed below are two schools of thought on how to measure transistor voltages in troubleshooting.

*Element to element.* Some troubleshooters prefer to measure transistor voltages from element to element (base to emitter, emitter to collector, collector to base) and note the *difference in voltage.* For example, in the circuit shown in Fig. 4-2, a 0.2 V differential exists between the base and emitter. The element-to-element method of measuring transistor voltages quickly establishes forward-bias or reverse-bias conditions.

*Element to ground.* The most common method of measuring transistor voltages is to measure from a common or ground to the element. CB service literature usually specifies transistor voltages this way. For example, all the voltages for the PNP shown in Fig. 4-2 are negative with respect to ground. (The positive test lead of the meter must be connected to ground, and the negative test lead is connected to each of the elements in turn.)

This method of labeling transistor voltages is sometimes confusing to those not familiar with transistors, because it appears to break the rules. (In a PNP transistor, some elements should be positive, but all elements are negative). However, the rules still apply.

In the case of the PNP shown in Fig. 4-2, the emitter is at $-0.3$ V, whereas the base is at $-0.5$ V. The base is *more negative* than the emitter. Thus, the emitter is *positive with respect to the base,* and the base-emitter junction is forward-biased (normal).

On the other hand, the base is at $-0.5$ V, whereas the collector is at $-7$ V. The base is *less negative* than the collector. Thus, the base is positive with respect to the collector, and the base-collector junction is reverse-biased (normally).

### 4-1.3 Troubleshooting with Transistor Voltages

This section presents an example of how voltages measured at the elements of a transistor may be used to analyze failure in solid-state CB circuits.

Sec. 4-1    Servicing Notes                                                    199

Assume that an NPN transistor circuit is measured and that the voltages found are similar to those shown in Fig. 4-3. Except in one case, these voltages indicate a defect. It is obvious that the transistor is not forward-biased because the base is less positive than the emitter (reverse bias for an NPN). The only CB circuit where this might be normal is one that requires a large trigger voltage or pulse (positive in this case) to turn it on (such as a squelch control circuit).

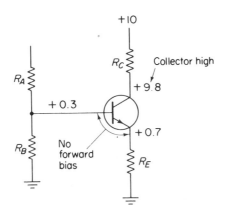

**Figure 4-3:** NPN transistor circuit with abnormal voltage (emitter base not forward-biased, collector voltage high)

The first troubleshooting clue in Fig. 4-3 is that the collector voltage is almost as large as the collector B+ voltage. This indicates that very little current is flowing through $R_C$ in the collector-emitter circuit. The transistor could be defective. However, the trouble is more likely to be caused by a problem in bias. The emitter voltage depends mostly on the current through $R_E$. Unless the value of $R_E$ has changed substantially (this would be unusual) the problem is one of incorrect bias on the base.

The next step in this case is to measure the bias-source voltage at $R_A$. If the bias-source voltage is (as shown in Fig. 4-4) at +0.7 V instead of the required 2 V, the problem is obvious; the external bias voltage is incorrect. This condition will probably show up as a defect in the power supply and will appear as an incorrect voltage in other circuits.

If the source voltage is correct, as shown in Fig. 4-5, the cause of the trouble is probably a defective $R_A$ or $R_B$ or a defect in the transistor.

The next step is to remove all voltage from the set and measure the resistance of $R_A$ and $R_B$. If either value is incorrect, the corresponding resistor must be replaced. If both values are correct, it is reasonable to check the value of $R_E$. However, it is more likely that the transistor is defective. This may be established by test and/or replacement.

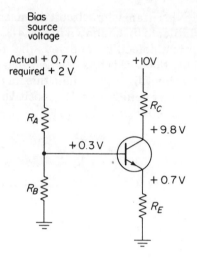

**Figure 4-4:** NPN transistor circuit with abnormal voltages (fault traced to incorrect bias source, bias voltage low)

*Practical in-circuit resistance measurements.* Do not attempt to measure resistor values in transistor circuits with the resistors still connected. Although this practice may be correct for vacuum-tube circuits, it is incorrect for transistor circuits, since it is possible that, for instance, the voltage produced by some ohmmeter batteries could damage some transistors. This does not mean that you cannot measure the

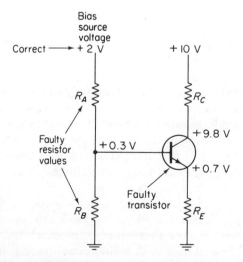

**Figure 4-5:** NPN transistor circuit with abnormal voltages (fault traced to bias resistors or transistors)

Sec. 4-1  Servicing Notes                                                    201

resistance values given in CB service literature resistance charts (if any), but you should not attempt to measure values of individual resistors without disconnecting the resistors (at least at one end).

The problem in measuring resistors in solid-state circuits is that transistor junctions will pass current in one direction and may complete a circuit through other resistors, producing a series or parallel combination and giving false indications. This situation may be prevented by disconnecting one resistor lead before making the resistance measurement.

For example, assume that an ohmmeter is connected across $R_B$ (Figs. 4-3 through 4-5) with the negative battery terminal of ohmmeter connected to ground, as shown in Fig. 4-6. Because $R_E$ is also connected to ground, the negative battery terminal is connected to the end of $R_E$.

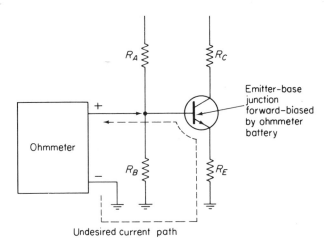

**Figure 4-6:** Example of in-circuit resistance measurements showing undesired current path through forward-biased transistor junction

Because the positive battery terminal is connected to the transistor base, the base-emitter junction is forward-biased, and there is electron flow. In effect, $R_E$ is now in parallel with $R_B$, and the ohmmeter reading is incorrect. This can be prevented by disconnecting either end of $R_B$ before making the measurement.

### 4-1.4  Testing Transistors in Circuit (Forward-bias Method)

Germanium transistors normally have voltage differential of 0.2 to 0.4 V between emitter and base; silicon transistors normally have a voltage differential of 0.4 to 0.8 V. The polarities of voltages at the emitter and base depend upon the type of transistor (NPN or PNP).

202                                    Ch. 4  BASIC CB SERVICE APPROACH

The voltage differential between emitter and base acts as a forward bias for the transistor. That is, a sufficient differential or forward bias will turn the transistor on, resulting in a corresponding amount of emitter-collector flow. Removal of the voltage differential or an insufficient differential will produce the opposite results. That is, the transistor is cut off (no emitter-collector flow or very little flow).

These forward-bias characteristics may be used to troubleshoot transistor circuits without removing the transistor and without using an in-circuit tester. The following discussions describe two methods of testing transistors in circuit; one by removing the forward bias and the other by introducing a forward bias.

*Removal of forward bias.* Figure 4-7 shows the test connections for an in-circuit transistor test by removal of forward bias. The procedure is simple: first, measure the emitter-collector differential voltage under

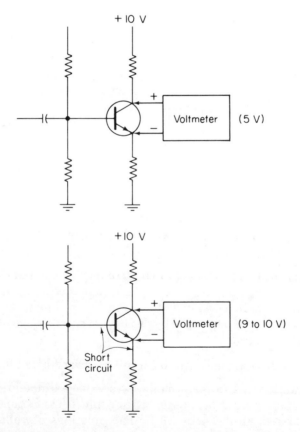

**Figure 4-7:** In-circuit transistor test (removal of forward bias)

Sec. 4-1 Servicing Notes

normal circuit conditions. Then, short the emitter-base junction and note any change in emitter-collector differential. (Do not short from base directly to ground.) If the transistor is operating, the removal of forward bias causes the emitter-collector current flow to stop, and the emitter-collector voltage differential increases. That is, the collector voltage rises to or near the power supply value.

For example, assume that the power supply voltage is 10 V and that the differential between the collector and emitter is 5 V when the transistor is operating normally (no short between emitter and base). When the emitter-base junction is shorted, the emitter-collector differential should rise to about 10 V (probably somewhere between 9 and 10 V).

*Application of forward bias.* Figure 4-8 shows the test connection for an in-circuit transistor test by the application of forward bias. The procedure is equally simple. First, measure the emitter-collector differential under normal circuit conditions. As an alternate, measure the voltage across $R_E$, as shown in Fig. 4-8.

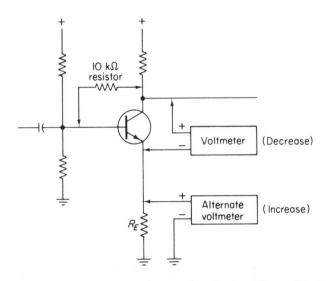

**Figure 4-8:** In-circuit transistor test (application of forward bias)

Next, connect a 10 kΩ resistor between the collector and base, as shown, and note any change in emitter-collector differential (or any change in voltage across $R_E$). If the transistor is operating, the application of forward bias will cause the emitter-collector current flow to start (or increase), and the emitter-collector voltage differential will decrease, or the voltage across $R_E$ will increase.

*Go/no-go test characteristics.* The test methods in Figs. 4-7 and 4-8 show that the transistor is operating on a go/no-go basis. This is usually sufficient for most dc and low-frequency ac applications. However, the tests do not show transistor gain or leakage, nor do they establish operation of the transistor at radio frequencies (RF).

The fact that these (or similar) in-circuit tests of a transistor do not establish all of the operating characteristics, particularly those that affect high-frequency RF circuits, raises a problem in CB troubleshooting. Some troubleshooters reason that the only satisfactory test of a transistor is in-circuit operation. If a transistor will not perform its function in a given circuit, the transistor must be replaced. Thus, the most logical method of test is replacement.

This reasoning is generally sound, but there is one exception. It is possible that a replacement transistor will not perform satisfactorily in critical circuits (high-frequency RF oscillator, mixer, or amplifier), even though the transistor is the correct type (and may even work in audio/modulator and dc switching circuits), thus creating a misleading situation. If such a replacement transistor does not restore the circuit to normal, the apparent fault is with another circuit, whereas the true cause of trouble is the new transistor. Fortunately, this does not happen often, even in critical circuits. However, you should be aware of the possibility in practical CB service work.

### 4-1.5 Using Transistor Testers in Troubleshooting

Transistors may be tested in or out of circuit using the commercial transistor testers described in Sec. 2-9. However, the value of their use in CB solid-state troubleshooting is generally a matter of opinion. At best, such testers show the gain and leakage of transistors at direct current or low frequencies under one set of conditions (fixed voltage, current, and so forth). For this reason, the use of transistor testers (and vacuum tube testers) is generally limited to the audio/modulator section components.

### 4-1.6 Testing Transistors out of Circuit

Four basic tests are required for transistors in practical troubleshooting: gain, leakage, breakdown, and switching time. All of these tests are best made with an oscilloscope using appropriate adapters (curve tracers, switching-characteristic checkers, and so forth). However, it is possible to test a transistor with an ohmmeter. These simple tests will show whether the transistor has leakage or shows some gain. As discussed, the only sure way to test a transistor is in the circuit in which the transistor is to be used.

Sec. 4-1 Servicing Notes

*Testing transistor leakage with an ohmmeter.* For test purposes (using an ohmmeter), transistors may be considered as two diodes connected back to back. Thus, each diode should show low forward resistance and high reverse resistance. These resistances can be measured with an ohmmeter, as shown in Fig. 4-9.

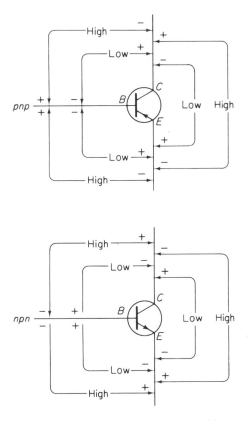

**Figure 4-9:** Transistor leakage tests with ohmmeter

The same ohmmeter range should be used for each pair of measurements (base to emitter, base to collector, and collector to emitter). However, avoid using the R×1 range or an ohmmeter with a high internal battery voltage. Either of these conditions can damage a low-power transistor.

If the reverse reading is low but not shorted, the transistor is leaking.

If both forward and reverse readings are very low or show a short, the transistor is shorted.

If both forward and reverse readings are very high, the transistor is open.

If the forward and reverse readings are the same or nearly equal, the transistor is defective.

A typical forward resistance is 300 to 700 Ω. However, a low-power transistor might show only a few ohms in the forward direction, especially at the collector-emitter junction. Typical reverse resistances are 10 to 70 kΩ.

The actual resistance values depend on the ohmmeter range and battery voltage; thus, the ratio of forward-to-reverse resistance is the best indicator. Almost any transistor will show a ratio of at least 30:1, and many transistors show ratios of 100:1 or greater.

*Testing transistor gain with an ohmmeter.* Normally, there will be little or no current flow between emitter and collector until the base-emitter junction is forward-biased. Thus, a basic gain test of a transistor can be made using an ohmmeter. The test circuit is shown in Fig. 4-10. In

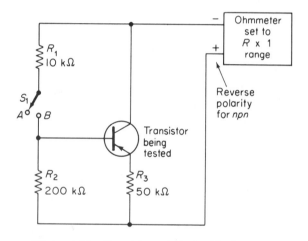

**Figure 4-10:** Transistor gain test with ohmmeter

this test, the R×1 range should be used. Any internal battery voltage can be used, provided it does not exceed the maximum collector-emitter breakdown voltage.

In position A of switch $S_1$, no voltage is applied to the base, and the base-emitter junction is not forward-biased. Thus, the ohmmeter reading should show a high resistance. When switch $S_1$ is set to B, the base-emitter circuit is forward-biased (by the voltage across $R_1$ and $R_2$), and current flows in the emitter-collector circuit. This is indicated by a lower resistance reading on the ohmmeter. A 10:1 resistance ratio is typical for an AF transistor.

Sec. 4-1 Servicing Notes 207

**4-1.7 Testing Diodes out of Circuit**

There are three basic diode tests. First, the diode must have the ability to pass current in one direction (forward current) and prevent or limit current flow (reverse current) in the opposite direction. Second, for a given reverse voltage, the reverse current should not exceed a given value. Third, for a given forward current, the voltage drop across the diode should not exceed a given value. If a diode is to be used in pulse or digital work (such as in some CB sets where the channel selection is through digital circuit operation), the switching time must also be tested. These tests are best performed using a scope with appropriate adapters.

However, because the elementary purpose of a diode is to prevent current flow in one direction while passing current in the opposite direction, a diode can be tested for basic CB troubleshooting purposes using an ohmmeter. In this case, the ohmmeter is used to measure forward and reverse resistance of the diode. The basic circuit is shown in Fig. 4-11.

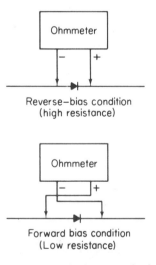

Figure 4-11: Basic diode test with ohmmeter

A good diode will show high resistance in the reverse direction and low resistance in the forward direction.

If resistance is low in the reverse direction, the diode is probably leaking.

If resistance is high in both directions, the diode is probably open.

A low resistance in both directions usually indicates a shorted diode.

It is possible for a defective diode to show a difference in forward and reverse resistance. The important factor in making a diode-resistance test is the ratio of forward-to-reverse resistance (often known as the front-to-back ratio or the back-to-front ratio). The actual ratio depends on the type of diode. However, as a guideline, a small-signal diode has a ratio of several hundred to one, whereas a power rectifier can operate satisfactorily with a ratio of 10:1.

### 4-1.8 Troubleshooting ICs

ICs are being used more frequently in modern CB set design to replace the audio/modulator section, the IF/detector section of the receiver, and in the PLL circuits. There is some difference of opinion on testing ICs in circuit or out of circuit during troubleshooting. An in-circuit test is the most convenient because the power source is available and you do not have to unsolder the IC. (Removal and replacement of an IC can be quite a job.)

Of course, first you must measure the dc voltages applied at the IC terminals to make sure that they are available and correct. If the voltages are absent or abnormal, this is a good starting point for troubleshooting.

With the power sources established, the in-circuit IC is tested by applying the appropriate input and monitoring the output. For example, if an IC replaces the audio/modulation section, the IC may be tested by injecting an audio signal at the IC input and listening for the tone on the loudspeaker. In some cases, it is not necessary to inject an input because the normal input is supplied by the circuits ahead of the IC.

One drawback in testing an IC in circuit is that the circuits before (input) and after (output) the IC may be defective. This can lead you to think that the IC is bad. For example, assume that the IC is used as the IF stages of the receiver section. To test such an IC, you inject a signal at the IC input and monitor the IC output. Now, assume that the IC output terminal is connected to a short circuit. No output will be indicated, even though the IC and the input signals are good. Of course, this will show up as an incorrect resistance measurement (if such measurements are made).

Out-of-circuit tests for ICs have two obvious disadvantages: you must remove the IC, and you must supply the required power. However, if you test a suspected IC after removal and find that it is operating properly out of circuit, it is logical to assume that there is trouble in the circuits connected to the IC. This is convenient to know before you go to the trouble of installing a replacement IC.

Sec. 4-1 Servicing Notes 209

*IC voltage measurements.* Although the test procedures for an IC are the same as those used for conventional transistor circuits of the same type, measurement of the static (dc) voltage applied to the IC is not. Some ICs used in CB require connection to both a positive and negative power source, although many can be operated from a single power supply source. Also, some ICs require equal power supply voltages (such as +9 V and −9 V). However, this is not the case with the circuits shown in Fig. 4-12, which requires +9 V at pin 8 and −3 V at pin 7.

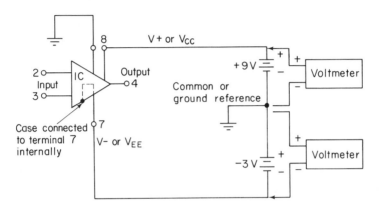

**Figure 4-12:** Measuring static (power source) voltages of ICs during troubleshooting

In most transistor circuits it is common to label one power supply lead positive and the other negative without specifying which (if either) is common or ground. However, in some ICs, it is necessary that all IC power supply voltages be referenced to a common or ground.

Manufacturers do not agree on power supply labeling for ICs. For example, one manufacturer might use V+ to indicate the positive voltage and V− to indicate the negative voltage. Another might use the symbols $V_{EE}$ and $V_{CC}$ to represent negative and positive, respectively. For this reason, you should study the schematic diagram carefully before measuring power source voltages during troubleshooting.

No matter what labeling is used, the IC may require two power sources, with the positive lead of one and the negative lead of the other tied to ground. When you work with such an IC, each voltage must be measured separately, as shown in Fig. 4-12.

Note that the IC case (such as a TO-5 type case) of the circuit shown in Fig. 4-12 is connected to pin 7. Such a connection is typical for most ICs (but not necessarily at pin 7). Thus, the case will be below ground (or hot) by 3 V.

### 4-1.9. Effects of Capacitors in Troubleshooting

During the troubleshooting process, suspected capacitors may be removed from the set and tested on capacitor checkers as discussed in Sec. 2-9, thus establishing that the capacitor value is correct. If the checker shows the value to be correct, it is reasonable to assume that the capacitor is not open, shorted, or leaking.

From another standpoint, a capacitor that shows no shorts, opens, or leakage, is assumed to be good. Thus, from a practical troubleshooting standpoint, a simple test that shows the possibility of shorts, opens, or leakage is usually sufficient.

There are two basic methods for a quick check of capacitors during troubleshooting. One method involves using the circuit voltages, while the other requires an ohmmeter.

*Checking capacitors with circuit voltages.* As shown in Fig. 4-13, this method involves disconnecting one lead of the capacitor (the ground or cold lead) and connecting a voltmeter between the disconnected lead and ground. In a good capacitor, there should be a momentary voltage indication (or surge) as the capacitor charges up the voltage at the hot end.

If the voltage indication remains high, the capacitor is probably shorted.

If the voltage indication is steady but not necessarily high, the capacitor is probably leaking.

If there is no voltage indication whatsoever, the capacitor is probably open.

*Checking capacitors with an ohmmeter.* As shown in Fig. 4-13, this method involves disconnecting one lead of the capacitor (usually the hot end) and connecting an ohmmeter across the capacitor. Make certain all power is removed from the circuit. As a precaution, short across the capacitor to make sure that no charge is being retained after the power is removed. In a good capacitor, there should be a momentary resistance indication (or surge) as the capacitor charges up to the voltage of the ohmmeter battery.

If the resistance indication is near zero and remains so, the capacitor is probably shorted.

If the resistance indication is steady at some high value, the capacitor is probably leaking.

If there is no resistance indication whatsoever, the capacitor is probably open.

Sec. 4-1 Servicing Notes

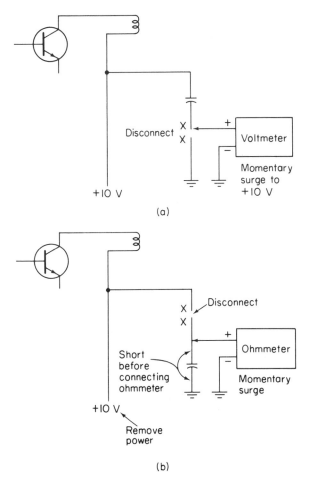

**Figure 4-13:** Check capacitors with circuit voltages (power applied) and with ohmmeter (power removed)

*Functions of capacitors in circuits.* The functions of capacitors in solid-state circuits are similar to the functions of those in vacuum-tube sets. However, the results produced by capacitor failure are not necessarily the same. An emitter-bypass capacitor is a good example.

The emitter resistor in a solid-state circuit (such as $R_4$ in Fig. 4-14) is used to stabilize the transistor's dc gain and prevent thermal runaway. With an emitter resistor in the circuit, any increase in collector current produces a greater drop in voltage across the resistor. When all other factors remain the same, the change in emitter voltage reduces the base-emitter forward-bias differential, thus tending to reduce collector

212  Ch. 4  BASIC CB SERVICE APPROACH

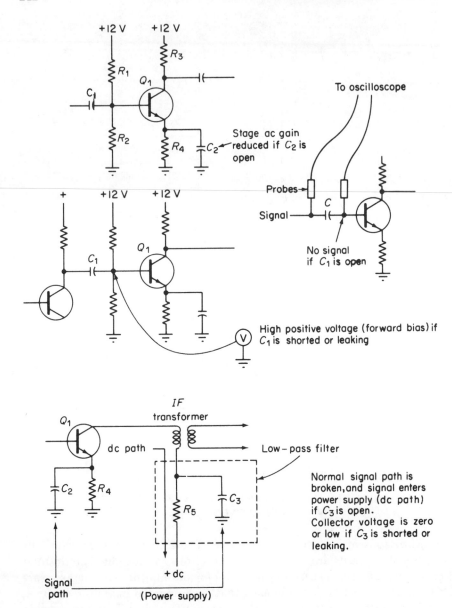

Figure 4-14: Effects of capacitor failure in solid-state circuits

current flow. When circuit stability is more important than gain, the emitter resistor is not bypassed. When ac or signal gain must be high (such as in the audio/modulation section input stages), the emitter resistance is bypassed to permit passage of the signal. If the emitter-bypass

## Sec. 4-1 Servicing Notes

capacitor is open, stage gain is reduced drastically, although the transistor's dc voltages remain substantially the same.

*Low-gain symptoms.* If there is a low-gain symptom in any solid-state amplifier with an emitter bypass, and the voltages appear normal, check the bypass capacitor. This may be done by substitution or by the procedures shown in Fig. 4-13.

*Coupling capacitors.* The functions of coupling (and decoupling) capacitors in solid-state sets are essentially the same as the functions of those in vacuum-tube sets. However, the capacitance values are much larger for solid-state, particularly at low frequencies such as those of the audio/modulation section. Electrolytic capacitors are usually necessary in solid-state audio/modulation circuits to achieve the large capacitance values required at audio frequencies. From a troubleshooting standpoint, electrolytics tend to have more leakage than mica or ceramic capacitors. However, good-quality electrolytics (typically the bantam type found in solid-state circuits) have leakages of less than 10 $\mu$A at normal operating voltage.

*Defects in coupling capacitors.* The function of $C_1$ in Fig. 4-14 is to pass signals from the previous stage to the base of $Q_1$. If $C_1$ is shorted or leaking badly, the voltage from the previous stage is applied to $Q_1$. This forward-biases $Q_1$, causing heavy current flow and possible burnout of the transistor. In any event, $Q_1$ is driven into saturation, and stage gain is reduced.

If $C_1$ is open, there is little or no change in the voltage at $Q_1$, but the signal from the previous stage will not appear at the base of $Q_1$. From a troubleshooting standpoint, a shorted or leaking $C_1$ will show up as abnormal voltages (and probably as distortion of the signal waveform). If $C_1$ is suspected of being shorted, open, or leaky, replace it (or check $C_1$ as shown in Fig. 4-13). An open $C_1$ will show up as a lack of signal at the base of $Q_1$, with a normal signal at the previous stage.

*Defects in decoupling or bypass capacitors.* The function of $C_3$ in Fig. 4-14 is to pass operating-signal frequencies to ground (to provide a return path) and to prevent signals from entering the power supply line or other circuits connected to the line. This decoupling or bypass function is most often found in receiver and transmitter circuits of the CB set. In effect, $C_3$ and $R_5$ form a low-pass filter that passes dc and very low frequency signals (well below the operating frequency of the circuit) through the power supply line. Higher-frequency signals are passed to ground and do not enter the power supply line.

If $C_3$ is shorted or leaking badly, the power supply voltage will be shorted to ground or greatly reduced. This reduction of collector voltage will make the stage totally inoperative or will reduce the output, depending on the amount of leakage in $C_3$.

If $C_3$ is open, there will be little or no change in the voltages at $Q_1$. However, the signals will appear in the power supply line. Also, signal gain will be reduced, and the signal waveform will be distorted. In some cases, at higher signal frequencies, the signal simply cannot pass through the power supply circuits. Because there is no path through an open $C_3$, the signal will not appear on the collector circuit. From a practical troubleshooting standpoint, the results of an open $C_3$ depend upon the values of $R_5$ (and the power supply components) as well as on the signal frequency involved.

### 4-1.10 Effects of Voltage on Circuit Resistance

The effects of shorts on resistors are less drastic in solid-state CB circuits than in vacuum-tube sets because of the lower voltage used in solid-state. For example, most solid-state CB sets operate with 12–13 V power sources. A 1 kΩ resistance shorted directly across a 12 V source produces only 12 mA current flow, or about 0.14 W. A one-quarter watt resistor can easily handle this power. On the other hand, the same resistance across a 300 V source (typical for vacuum-tube circuits) produces about 0.3 A current flow, or about 90 W. This will destroy all but heavy power transistors.

For these reasons, resistors do not burn out as often in solid-state CB as they do in similar vacuum-tube sets. Also, solid-state resistance values do not usually change as a result of prolonged heating. There are exceptions, of course, but most solid-state CB circuit troubles are the result of defects in capacitors, transistors, or diodes, in that order.

### 4-1.11 Effects of Voltage on Poor Solder Joints

The low voltages in solid-state CB have just the opposite effect of high voltages on poor solder joints (so-called cold solder joints) and partial breaks in printed wiring. The high voltages in vacuum-tube sets can often overcome the resistance created by cold solder joints and partial printed-circuit breaks.

When there is no obvious cause for a low voltage at some point in the circuit, or there is an abnormally high resistance, look for cold solder joints or defects in printed-circuit wiring. Use a magnifying glass to locate defects in printed wiring. Minor breaks in printed wiring can sometimes be repaired by applying solder at the break; however, this is recommended only as a temporary measure. Under emergency condi-

Sec. 4-2 Frequency Synthesizer, Oscillator, and PLL Troubleshooting 215

tions, it is possible to run a wire between two points on either side of the break, but it is recommended that the entire board be replaced as soon as practical.

*Finding cold solder joints.* Cold solder joints may sometimes be found with an ohmmeter. Remove all power and connect the ohmmeter across two wires leading out of the suspected cold solder joint, as shown in Fig. 4-15. Flex the wires by applying pressure with the ohmmeter prod tips. Switch the ohmmeter to different ranges, and check for any change in resistance. For example, a cold solder joint may appear to be good on the high ohmmeter ranges but open on the lower ranges. Look for resistance indications that tend to drift or change when the ohmmeter is returned to a particular scale. If a cold solder joint is suspected, reheat the joint with a soldering tool; then recheck the resistance.

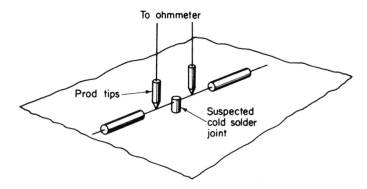

**Figure 4-15:** Locating cold solder joints with ohmmeter

## 4-2 FREQUENCY SYNTHESIZER, OSCILLATOR, AND PLL TROUBLESHOOTING

As noted in the discussion on circuits in Chapter 3, the frequency control circuits of a CB set can be quite simple or fairly complex, depending on the set. In the simplest form, such as a hand-held CB, there are two oscillators, one for transmitter frequency control and one for the receiver local oscillator. In a "typical" set there will be two or three oscillators, combined with one or more mixers, to form a frequency synthesizer that produces transmitter and receiver local oscillator signals. A PLL set contains a standard oscillator (probably within the PLL IC, but having an external crystal) and one or two other oscillators to form the complete frequency synthesizer.

No matter how complex the circuits appear, *they are essentially oscillators*, and can be treated as such from a troubleshooting standpoint. That is, each circuit contains oscillators, which produce crystal-controlled signals. These signals must have a given amplitude and must be at a given frequency for the set to operate properly. Thus, if you measure the signals and find them to be of the correct frequency and amplitude, the oscillators are good. Leave them alone. If either or both amplitude and frequency are incorrect, the oscillator is bad, and the problem is localized.

### 4-2.1 Oscillator Troubleshooting Problems

The first step in checking any oscillator circuit is to measure both the amplitude and frequency of the output signal. Most CB set oscillators have a built-in test point. If not, the signal may be monitored at the collector or emitter (plate or cathode for vacuum tube). Signal amplitude may be measured with a meter (Sec. 2-4) or an oscilloscope (Sec. 2-3). Both instruments will require an RF probe (Sec. 2-5). The simplest way to measure oscillator signal frequency is with a frequency counter (Sec. 2-6).

*Oscillator frequency problems.* When you measure the oscillator signal, the frequency will be: (1) right on, (2) slightly off, or (3) way off.

If the frequency is slightly off, it is possible to correct the problem with adjustment. Most CB oscillators are adjustable; usually the RF coil or transformer is slug-tuned. The most precise adjustment is obtained by monitoring the oscillator signal with a frequency counter and adjusting the circuit for exact frequency. However, it is also possible to adjust an oscillator using a meter or scope. When the circuit is adjusted for *maximum signal amplitude,* the oscillator is at the crystal frequency. However, it is possible (but not likely) that the oscillator is being tuned to a harmonic (multiple or submultiple) of the crystal frequency. The frequency counter will show this, whereas the meter or scope will not.

If oscillator frequency is way off, look for a defect rather than improper adjustment. For example, the coil or transformer may have shorted turns, the transistor or capacitor may be leaking badly, or the *wrong crystal* was installed in the right socket (this does happen).

*Oscillator signal amplitude problems.* When you measure the oscillator signal, the amplitude will be: (1) right on, (2) slightly low, or (3) very low.

If the amplitude is slightly low, it is possible to correct the problem with adjustment. Monitor the signal with a meter or scope, and adjust the oscillator for maximum signal amplitude. This will also lock the oscillator on the correct frequency.

Sec. 4-2 Frequency Synthesizer, Oscillator, and PLL Troubleshooting 217

If the amplitude is very low, look for defects such as low power supply voltages, leaking transistor and/or capacitors, and shorted coil or transformer turns. Usually, when signal output is very low, there will be other indications such as abnormal voltage and resistance values.

*Oscillator bias problems.* One of the problems in troubleshooting solid-state oscillator circuits is the bias arrangement. RF oscillators are generally reverse-biased, so that they conduct on half-cycles. However, the transistor is initially forward-biased by dc voltages (through the bias network as shown in Fig. 4-16). This turns the transistor on so that the collector circuit starts to conduct. Feedback occurs, and the transistor is driven into heavy conduction.

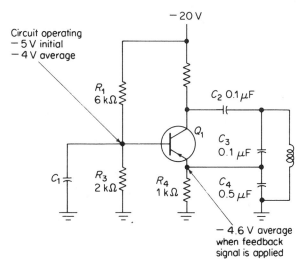

**Figure 4-16:** Class C RF oscillator (reverse-biased or zero-biased with circuit operating)

During the time of heavy conduction, a capacitor connected to the transistor base is charged in the forward-bias direction. When saturation is reached, there is no further feedback, and the capacitor discharges. This reverse-biases the transistor and maintains the reverse bias until the capacitor has discharged to a point where the fixed forward bias again causes conduction.

This condition presents a problem in the operation of class C solid-state RF oscillators. If the capacitor is too large, it may not discharge in time for the next half-cycle. In that case, the class C oscillator acts as a blocking oscillator, controlling the frequency by the capacitance and resistance of the circuit. If the capacitor is too small, the class C oscillator may not start at all. This same condition is true if the capacitor is *leaking badly.* From a practical troubleshooting standpoint, the measured condition of bias on a solid-state oscillator can provide a good clue to operation, if you know how the oscillator is supposed to operate.

The oscillator in Fig. 4-16 is initially forward-biased through $R_1$ and $R_3$. As $Q_1$ starts to conduct and in-phase feedback is applied to the emitter (to sustain oscillation), capacitor $C_1$ starts to charge. When saturation is reached (or approached) and the feedback stops, capacitor $C_1$ then discharges in the opposite polarity, reverse-biasing $Q_1$. The value of $C_1$ is selected so that $C_1$ discharges to a voltage less than the fixed forward bias before the next half-cycle. Thus, transistor $Q_1$ conducts on slightly less than the full half-cycle. Typically, a class C RF oscillator such as the one shown in Fig. 4-16 conducts on about 140° of the 180° half-cycle.

Exploring the subject of bias further, it is commonly assumed that transistor junctions (and diodes) start to conduct as soon as forward voltage is applied; this is not true. Figure 4-17 shows characteristic curves for three different types of transistor junctions. All three junctions are silicon, but the same condition exists for germanium junctions. None of the junctions conduct noticeably at 0.6 V, but current starts to rise at that point. At 0.8 V, one junction draws almost 80 mA. At 1 V, the dc resistance is on the order of 2 or 3 Ω, and the transistor draws almost 1 A. In a germanium transistor, noticeable current flow occurs at about 0.3 V.

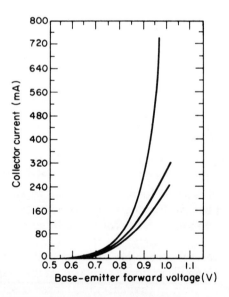

**Figure 4-17:** Characteristic curves for silicon transistor junctions (collector current flow versus base-emitter forward voltage)

For troubleshooting purposes, bias measurements provide a clue to the performance of solid-state oscillators, although such measurements do not provide positive proof. The one sure test of an oscillator is to measure output signal amplitude and frequency.

*Oscillator quick-check.* It is possible to check whether an oscillator circuit is oscillating by using a voltmeter and large value capacitor (typically 0.01 or larger). Measure either the collector or emitter voltage with the oscillator operating normally, and then connect the capacitor from base to ground. This should stop oscillation, and the emitter or collector voltage will change. When the capacitor is removed, the voltage will return to normal. If there is no change when the capacitor is connected, the oscillator is probably not oscillating. In some oscillators, you will get better results by connecting the capacitor from the collector to ground. Also, do not expect the voltage to change on an element without a load. For example, if the collector is connected directly to B+, this voltage will not change, with or without oscillation.

## 4-3 AUDIO/MODULATION SECTION TROUBLESHOOTING

The audio/modulation section of a CB set is essentially an audio amplifier. For that reason, we shall discuss the basic troubleshooting approach for amplifiers, the test procedures normally associated with amplifier troubleshooting, and some practical notes on the analysis of basic amplifier circuits.

### 4-3.1 Basic Amplifier Troubleshooting Approach

The basic troubleshooting approach for an amplifier involves signal tracing. The input and output signals or waveforms may be monitored on a voltmeter or oscilloscope. Any stage showing an abnormal signal (in amplitude, waveshape, and so on) or the absence of an output signal with a known good input signal points to a defect in that stage. Voltage and resistance measurements on all elements of the vacuum tube or transistor will then pinpoint the problem.

Some CB technicians feel that the oscilloscope is the most logical instrument for checking audio/modulation circuits. The scope can duplicate every function of an electronic voltmeter in troubleshooting. In addition, the scope offers the advantage of a visual display for common audio-amplifier conditions such as distortion, hum, noise, ripple, and oscillation.

When troubleshooting amplifier circuits using signal tracing, the scope and meter are used in much the same way. A signal is introduced into the input by a signal generator, as shown in Fig. 4-18. The amplitude (and waveform if a scope is used) of the input signal are measured. The scope or meter probe is then moved to the input and output of each stage in turn until the final output is reached. The gain of each stage is measured as a voltage change from input to output. If a scope is used, it is also possible to observe any change in waveform from that applied to the input. Thus, stage gain and distortion (if any) are established quickly with a scope.

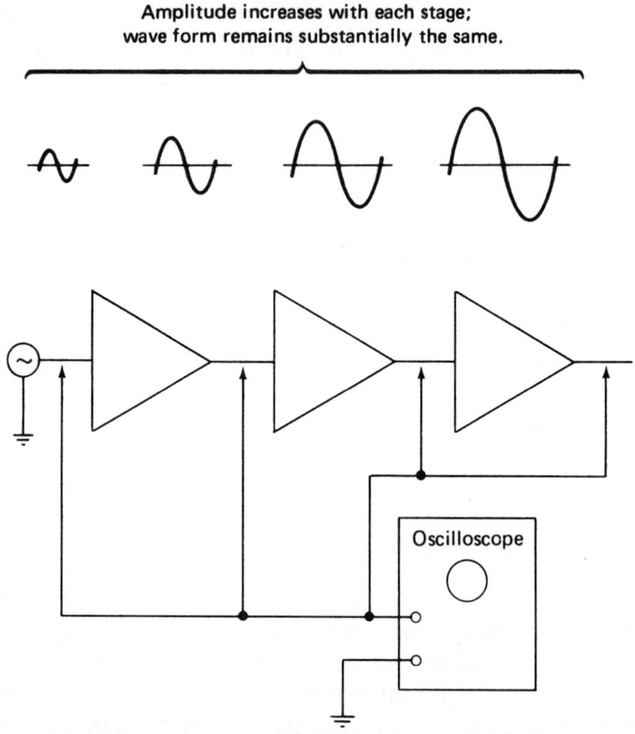

**Figure 4-18:** Basic signal-tracing trouble amplifier circuit using sine waves and an oscilloscope

### 4-3.2 Amplifier Frequency Response

It is not generally necessary to run a full frequency response check when servicing the audio/modulation section, except possibly if audio distortion occurs during transmission and reception. In any event, you should understand the basic frequency response measurement procedure.

## Sec. 4-3 Audio/Modulation Section Troubleshooting

The frequency response of the audio/modulation section may be measured with an audio signal generator and a meter or scope. The signal generator is tuned to various frequencies, and the resulting output response is measured at each frequency. The results are then plotted in the form of a graph or *response curve*, as shown in Fig. 4-19.

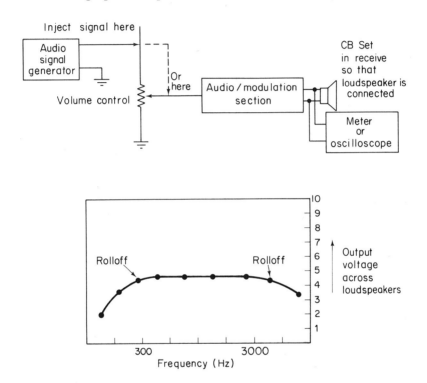

**Figure 4-19:** Audio/modulation section frequency-response test connections and typical response curve

The signal generator must produce a *constant amplitude* at all frequencies. The generator signal is varied in frequency (but not in amplitude) across the entire operating range of the audio/modulation section. Typically, this is in the 300 to 3,000 Hz range, although most sets will pass signals of lower frequency. The voltage output at various frequencies across the range is plotted on a graph as follows:

1. Connect the equipment as shown in Fig. 4-19. The set should be in the receive mode so that the loudspeaker is connected at the output of the audio/modulation section.

2. Initially, set the generator frequency to the low end of the range. Then set the generator amplitude to the desired input level. Typically, the volume control

(input to the audio/modulation section) receives about 1 V or less of audio from the receiver detector, with a normal (1 to 30 $\mu$V) signal at the receiver antenna input. Thus, the test should be made with 1 V or less from the generator.

3. In the absence of a realistic test input voltage, set the generator output to an arbitrary value. A simple method of finding a satisfactory input level is to monitor the circuit output at the loudspeaker (with a meter or scope) and increase the generator output at 1 kHz until the circuit is overdriven. This point is indicated when further increases in the generator output do not cause further increases in meter reading (or the output waveform peaks begin to flatten on the scope display). Set the generator output *just below* this point and measure the generator voltage. The generator output will probably be in the order of 100 to 500 mV. Keep the generator at this voltage throughout the test.

4. Set the volume control to midrange or some other arbitrary point when making the initial frequency response measurement. The response measurements can then be repeated at different volume control setting. If the audio/modulation section has any other controls, set these to midrange or at their typical operating position.

5. Record the circuit output voltage (at the loudspeaker) on the graph. Without changing the generator output amplitude, increase the generator frequency by some fixed amount, and record the new circuit output voltage. The amount of frequency increase between each measurement is an arbitrary matter. Use an increase of 10 Hz where rolloff occurs and 100 Hz at the middle frequencies.

6. Repeat the process, checking and recording the output voltage at each of the checkpoints in order to obtain a frequency response curve. For a typical audio/modulation section, the curve will resemble that shown in Fig. 4-19, with a flat portion across the middle frequencies and a rolloff at each end.

7. After the initial frequency response check, the effects of the volume control should be checked. The volume control should have the same effect all across the frequency range.

8. Note that generator output *may* vary with changes in frequency, a fact often overlooked in making a frequency response test during troubleshooting. Even precision laboratory generators may vary in output with changes in frequency, resulting in considerable error. It is recommended that the generator output be monitored after each change in frequency (some audio generators have a built-in output meter). Then, if necessary, the generator output amplitude can be reset to the correct value. It is more important that the generator output amplitude remain *constant*, rather than set at some specific value, when making a frequency response check.

### 4-3.3 Audio Modulation Section Voltage Gain

Voltage gain in the audio/modulation section is measured in the same way as frequency response. The ratio of output voltage to input voltage (at any given frequency or across the entire frequency range) is the voltage gain. Because the input voltage (generator output) is held

Sec. 4-3  Audio/Modulation Section Troubleshooting    223

constant for a frequency response test, a voltage gain curve should be identical to a frequency response curve. For example, assume that the generator output is 100 mV, and the voltage across the loudspeaker is 3 V. This means that the voltage gain is 30 (3 V/100 mV).

### 4-3.4 Power Output and Gain Measurements

The power output of the audio/modulation section is found by noting the output voltage across the loudspeaker at any frequency or across the entire frequency range. Power output is found by: output voltage$^2$/speaker impedance. For example, if the output voltage is 3 V and the speaker impedance is 4 $\Omega$, the power output is 2.25 W ($3^2/4$).

To find the power gain of the audio/modulation section, it is necessary to determine both the input and output power. Input power is found in the same way as output power except that the input impedance must be known (or calculated). For a test such as that shown in Fig. 4-19, assume that the input impedance is equal to the volume control resistance. With the input power known (or estimated), the power gain is the ratio of output power to input power. For example, assume an output power of 2.25 W. with an 100 mV generator signal across a 10 k$\Omega$ volume control. This is an input power of 1 $\mu$W with an output of 2.25 W, or a power gain of 2,250,000.

### 4-3.5 Checking Distortion Sine-wave Analysis

All amplifiers are subject to distortion; that is, the output signal may not be identical to the input signal. Theoretically, the output should be identical to the input except for amplitude. Some troubleshooting techniques are based on analyzing the waveshape of signals passing through an amplifier to determine possible distortion. If distortion (or an abnormal amount of distortion) is present, the circuit is further checked using the usual troubleshooting methods (localization, voltage measurements, etc.).

Amplifier distortion may be checked by sine-wave analysis. The procedures are the same as those used for signal tracing (Sec. 4-3.1 and Fig. 4-18). However, the primary concern in distortion analysis is deviation of the amplifier output waveform from the input waveform. If there is no change (except in amplitude), there is no distortion. If there is a change in the waveform, the nature of the change often reveals the cause of distortion. For example, the presence of second or third harmonics often distorts the fundamental.

In practical troubleshooting, analyzing sine waves to pinpoint amplifier problems that produce distortion is a difficult job that requires considerable experience. Unless the distortion is severe, it may pass unnoticed. Sine waves are best used where *harmonic-distortion* or *intermodulation-distortion* meters are combined with the scope for distortion analysis. Distortion meters are generally of little value in CB service work, although they are quite useful in establishing the characteristics of hi-fi and stereo amplifiers. If a scope is used alone, square waves provide the best basis for distortion analysis. (The reverse is true for frequency response and power measurements.)

### 4-3.6 Checking Distortion by Square-wave Analysis

The procedure for checking distortion by means of square waves is essentially the same as that used with sine waves. Distortion analysis is more effective with square waves because of their high odd-harmonic content and because it is easier to see a deviation from a straight line with sharp corners than from a curving line.

Square waves are introduced into the circuit input, and the output is monitored with a scope, as shown in Fig. 4-20. The primary concern is deviation of the amplifier output waveform from the input waveform (which is also monitored on the scope). If the scope has the *dual-trace* feature, the input and output may be monitored simultaneously. If there is a change in waveform, the nature of the change often reveals the cause of distortion.

The third, fifth, seventh, and ninth harmonics of a clean square wave are emphasized. If an amplifier passes a given frequency and produces a clean square-wave output, it is safe to assume that the frequency response is good up to at least nine times the square-wave frequency. Thus, if the square-wave generator is set to 300 Hz and produces good response, the audio/modulation section may be considered to be good across the normal frequency range.

### 4-3.7 Background Noise Measurement

If the vertical channel of a scope is sufficiently sensitive, the scope may be used to check and measure the background noise level of the audio/modulation section, as well as to check for the presence of hum, oscillation, and the like. The scope's vertical channel should be capable of a measurable deflection with about 1 mV (or less), because this is the background noise level of many amplifiers.

The basic procedure consists of measuring amplifier output with the volume control at maximum but without an input signal. The oscilloscope is superior to a voltmeter for noise level measurement because the frequency and nature of the noise (or other signal) are displayed visually.

## Sec. 4-3 Audio/Modulation Section Troubleshooting

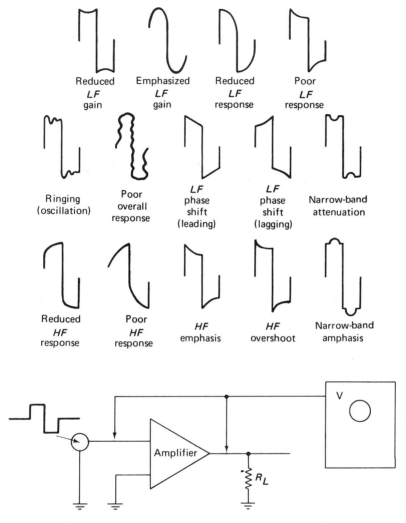

**Figure 4-20:** Amplifier square-wave distortion analysis

The basic connections for measuring the level of background noise are shown in Fig. 4-21. The scope gain is increased until there is a noise or "hash" indication.

A noise indication may be caused by pickup in the oscilloscope leads. If in doubt, disconnect the oscilloscope leads from the loudspeaker and see if the noise indication remains.

If the set is operated from line power (such as a base station), and it is suspected that there is a 60 Hz line hum present in the amplifier output (picked up from the power supply or any other source), set the scope's

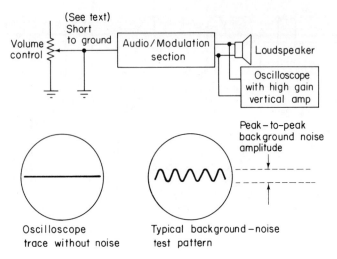

**Figure 4-21:** Audio/modulation section background-noise test connections

SYNC control (or whatever other control is required to synchronize the scope trace at the line frequency) to LINE. If a stationary signal pattern appears, it is the result of the line hum.

If a signal appears that is not at the line frequency, it may be the result of oscillation in the amplifier or stray pickup. Short across the volume control; if the signal remains, it is probably oscillation in the amplifier.

### 4-3.8 Feedback Amplifier Troubleshooting

Troubleshooting amplifiers without feedback is a relatively simple procedure. When the amplifier has feedback, the task is more difficult. Problems such as measurement of gain can be of particular concern.

For example, if you try opening the loop to make gain measurements, you will usually find so much gain that the amplifier saturates and the measurements are meaningless. On the other hand, if you start making waveform measurements on a working closed-loop system, you will often find the input and output signals are normal (or near normal), although many of the waveforms are distorted inside the loop. For this reason, feedback loops, especially internal-stage feedback loops, require special attention.

In any amplifier there are three basic causes of distortion: overdriving, operating the transistor at the wrong bias point, and the inherent nonlinearity of any solid-state device.

*Overdriving* may be caused by many conditions (too much input signal, such as results from yelling into the microphone, too much gain in the previous stage, volume control set too high, etc.). The net result is

## Sec. 4-3 Audio/Modulation Section Troubleshooting

that the output signal is clipped on one peak because the transistor is driven into saturation and on the other peak because the transistor is driven below cutoff.

Operating at the *wrong bias point* may also produce clipping, but of only one peak. For example, if the input signal is 1 V and the transistor is biased at 1 V, the input will swing from 0.5 to 1.5 V. Assume that the transistor saturates at any point where the base goes above 1.6 V and is cut off when the base goes below 0.4 V. No problem occurs when the bias is correct at 1 V.

But now assume that the bias point is shifted (because of component aging, transistor leakage, etc.) to 1.3 V. When the 1 V input signal is applied, the base swings from 0.8 to 1.8 V, and the transistor saturates when one peak goes from 1.6 to 1.8 V. If, on the other hand, the bias point is shifted to 0.7 V, the base swings from 0.2 to 1.2 V, and the opposite peak is clipped as the transistor goes into cutoff.

Even if the transistor is not overdriven, it is still possible to operate a transistor on a *nonlinear portion* of its curve because of wrong bias. Some portion of the input-output curve of all transistors is more linear than other portions. That is, the output increases (or decreases) directly in proportion to input. An increase of 10% at the input produces an increase of 10% at the output. Ideally, transistors are operated at the center of the linear curve. If the bias point is changed, the transistor can operate on a portion of the curve that is less linear than the desired point.

The *inherent nonlinearity* of any solid-state device (diode, transistor, etc.) may produce distortion even if a stage is not overdriven and is properly biased; that is, the output never increases (or decreases) in proportion to the input. For example, an increase of 10% at the input may produce an increase of 13% at the output. This is one of the main reasons for feedback in amplifiers where low distortion is required.

In summary, a negative-feedback loop operates to minimize distortion, in addition to stabilizing gain. The feedback takeoff point has the least distortion of any point within the loop. From a practical troubleshooting standpoint, if the final output distortion and the overall gain are within limits, all the stages within the loop may be considered to be operating properly. Even if there is some abnormal gain in one or more of the stages, the overall feedback system has compensated for the problem. Of course, if the overall gain and/or distortion are not within limits, the individual stages must be checked.

Most feedback amplifier problems may be pinpointed by waveform and voltage measurements, as discussed throughout this book. You should give special attention to the following paragraphs when you are troubleshooting any audio/modulation section containing a feedback network.

*Opening the loop.* Some service literature recommends that the loop be opened and the circuits checked under no-feedback conditions. In some cases, this procedure can cause circuit damage. But even if there is no damage, the technique is rarely effective. Open-loop gain is usually so great that some stage will block or distort badly. If the technique is used, as it must be for some circuits, keep in mind that distortion is increased when the loop is opened. That is, a normally closed-loop amplifier may show considerable distortion when operated as an open-loop device, even though the amplifier is good.

*Measuring stage gain.* Care should be taken when measuring the gain of stages in a feedback amplifier. For example, the base-to-ground voltage is not always the same as the input voltage. To get the correct value, connect the low side of the measuring device (ac voltmeter or scope) to the emitter, and connect the other lead (high side) to the base, as shown in Fig. 4-22. In effect, measure the signal that appears across the base-emitter junction. This measurement will include the effect of the feedback signal.

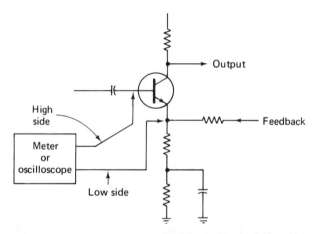

**Figure 4-22:** Measuring input-signal voltage or waveforms

As a general safety precaution, never connect the ground lead of a voltmeter or scope to the base of a transistor unless the lead connects back to an isolated inner chassis on the meter or scope. The reason for this precaution is that large ac ground-loop currents (between the measuring device and the set being serviced) may flow through the base-emitter junction and possibly burn out the transistor.

*Low-gain problems.* Low gain in a feedback amplifier may also result in distortion. That is, if gain is normal in the feedback amplifier,

## Sec. 4-3 Audio/Modulation Section Troubleshooting

some distortion can be overcome. With low gain, the feedback may not be able to bring the distortion within limits. Of course, low gain by itself is sufficient cause to troubleshoot an amplifier (with or without feedback).

Assume, for example, the classic failure pattern of a solid-state audio/modulation section that had been working properly but that now shows a decrease in output of about 10%. It is necessary to advance the volume control both during receive (to hear the audio) and transmit (to get proper modulation). This indicates a general deterioration of performance rather than a major breakdown.

In troubleshooting such a situation, if waveforms indicate low gain and element voltages are normal, you should try replacing the transistors. Of course, you must never overlook the possibility of open or badly leaking emitter-bypass capacitors. If the capacitors are open or leaking (acting as a resistance in parallel with the emitter resistor), there will be considerable negative feedback and little ac gain. Remember, a completely shorted emitter-bypass capacitor produces an abnormal dc voltage indication at the transistor emitter.

*Distortion problems.* As previously discussed, distortion may be caused by improper bias, overdriving (too much gain), or underdriving (too little gain, preventing the feedback signal from countering the distortion). One problem often overlooked in an audio/modulation section that shows a pattern of distortion trouble is overdriving as the result of transistor leakage. (The problem of transistor leakage is discussed further in Sec. 4-3.9.)

Generally, it is assumed that the collector-base leakage will reduce gain, because the leakage is in opposition to the signal-current flow. Although this is true in the case of a single stage, it may not be true when more than one feedback stage is involved.

Whenever there is collector-base leakage, the base assumes a voltage nearer to that of the collector (nearer than is the case without leakage). This increases both transistor forward bias and transistor current flow. An increase in the transistor current causes a lower $h_{ib}$ (ac input resistance, grounded-base configuration), which in turn causes a reduction in common-emitter input resistance; this reduction may or may not cause a gain reduction (depending on where the transistor is located in the amplifier). If the feedback amplifier is *direct-coupled,* the effects of feedback are increased, since the operating point (base bias) of the following stage is changed, possibly resulting in distortion.

### 4-3.9 Effects of Transistor Leakage on Amplifier Gain

When there is considerable transistor leakage in a solid-state audio/modulation section, the gain is reduced to zero and/or the signal is drastically distorted. Such a condition also produces abnormal waveforms and transistor voltages. These indications make troubleshooting easy, or at least relatively easy. The troubleshooting problem becomes really difficult when there is just enough leakage to reduce amplifier gain, but not enough leakage to distort the waveform seriously or produce transistor voltages that are way off.

Collector-base leakage is the most common form of transistor leakage and produces a classic condition of low gain (in a single stage). When there is any collector-base leakage, the transistor is forward-biased, or the forward bias is increased, as shown in Fig. 4-23.

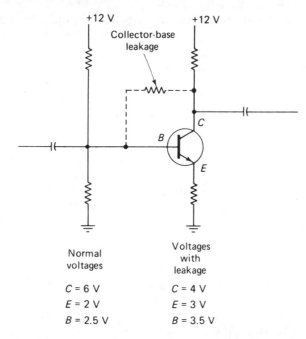

**Figure 4-23:** Effect of collector-base leakage on transistor element voltages

Collector-base leakage has the same effect as a resistance between the collector and base. The base assumes the same polarity as the collector (although at a lower value), and the transistor is forward-biased. If leakage is sufficient, the forward bias may be enough to drive the transistor into or near saturation. When a transistor is operated at or near the saturation point, the gain is reduced (for a single stage), as shown in Fig. 4-24.

If the normal transistor element voltages are known (from the service

Sec. 4-3  Audio/Modulation Section Troubleshooting            231

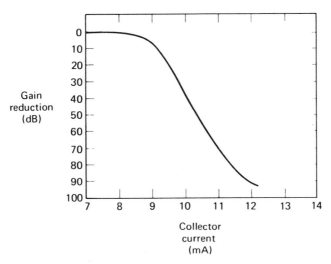

**Figure 4-24:** Relative gain of solid-state amplifier at various average collector-current levels

literature or from previous readings taken when the amplifier was operating properly), excessive transistor leakage may be spotted easily because all the transistor voltages will be off. For example, in Fig. 4-23, the base and emitter will be high and the collector will be low (when measured in reference to ground).

If the normal operating voltages are not known, the transistor may seem to be good because all the voltage relationships are normal. That is, the collector-base junction is reverse-biased (collector more positive than base for an NPN), and the base-emitter junction is forward-biased (emitter less positive than base for an NPN).

A simple way to check transistor leakage is shown in Fig. 4-25. Measure the collector voltage to ground. Then short the base to the emitter and remeasure the collector voltage. If the transistor is not leaking, the base-emitter short will turn the transistor off, and the collector voltage will rise to the same value as the supply (or nearly so). If there is any leakage, a current path will remain (through the emitter resistor, emitter-base short, collector-base leakage path, and collector resistor). There will be some voltage drop across the collector resistor, and the collector will have a voltage at some value lower than the supply.

Note that most meters draw current, and this current passes through the collector resistor. This can lead to some confusion, particularly if the meter draws heavy current (has a low ohms-per-volt rating). To eliminate any doubt, connect the meter to the supply through a resistor with the *same* value as the collector resistor. The drop, if any, should be the

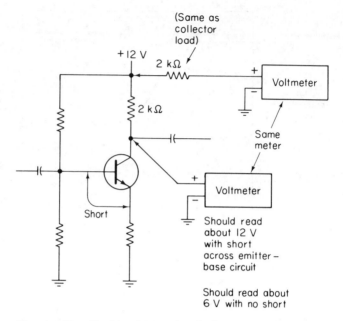

**Figure 4-25:** Checking for transistor leakage in amplifier circuit

same as it is when the transistor is measured to ground. If the drop is much different (lower) when the collector is measured, the transistor is leaking.

For example, assume that in the circuit of Fig. 4-25 the supply is 12 V, the collector resistance is 2 kΩ, and the collector measures 4 V with respect to ground. This means that there is an 8 V drop across the collector resistor and a collector current of 4 mA (8/2,000=4 mA). Normally, the collector is operated at about one-half the supply voltage (in this case, 6 V). However, simply because the collector is at 4 V instead of 6 V does not make the circuit faulty; some circuits are designed that way.

In any event, the transistor should be checked for leakage using the emitter-base short test shown in Fig. 4-25. Now, assume that the collector voltage rises to 10.5 V when the base and emitter are shorted. This indicates that the transistor is cutting off but that there is still some current flow through the collector resistor, about 1 mA (2/2,000=1 mA).

A current flow of 1 mA is high for a meter. However, to confirm a leaking transistor, connect the same meter through a 2 kΩ resistor (the same as the collector load) to the 12 V supply (preferably at the same point where the collector resistor connects to the power supply). Now, assume that the indication is 11.7 V through the external resistor. This indicates that there is some transistor leakage.

Sec. 4-3  Audio/Modulation Section Troubleshooting         233

The amount of transistor leakage may be estimated as follows: 11.7−10.5=1.2 V drop, and 1.2/2,000=0.6 mA. However, from a practical troubleshooting standpoint, the presence of any current flow with the transistor supposedly cut off is sufficient reason to replace the transistor.

### 4-3.10  Example of Audio/Modulation Section Troubleshooting

This step-by-step troubleshooting problem involves locating the defective part in a solid-state audio/modulation section, and then repairing the trouble. The problem has been localized to the audio/modulation section using the basic procedures in Chapter 1 and Fig. 1-7. The trouble symptoms are "no modulation but the carrier is present as indicated by the RF meter; there is no sound in the speaker with the volume control full on."

*General instructions.* The schematic diagram for the audio/modulation section is shown in Fig. 4-26. No test points (as such) are given on the schematic, and the voltage information is incomplete. There is no resistance information. The only service information availa-

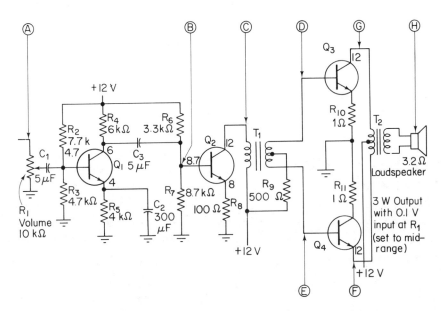

**Figure 4-26:** Schematic diagram with test points of audio/modulation section

ble is the schematic and a note (on the schematic) stating that the output is 3 W across a 3.2 Ω loudspeaker when a 0.1 V (100 mV) signal is introduced at the input (across volume control $R_1$) and $R_1$ is at midrange.

Using this fragmentary information (which is probably more than you will get in many practical CB service situations) you can pencil in the logical test points, as shown in Fig. 4-26. The test points are logical because they show the input and output of each stage. Note that the test points show linear, separating, and meeting paths (as discussed in Chapter 1).

Also, using what you know, the input to be introduced at test point A should be 0.1 V at some audio frequency. Thus, you can connect an audio generator to test point A (as shown in Fig. 4-27) and set the generator to produce 0.1 V (100 mV) at a frequency of 1,000 Hz. Under these conditions, the output voltage at test point H should be about 3.1 V. How do you know this? The output is supposed to be 3 W across a 3.2 Ω loudspeaker, and $E=\sqrt{PR}$, or $E=\sqrt{3\times3.2}=\sqrt{9.6}=3.1$ V.

You have no idea what the signals at other test points are. But you know that (with an appropriate signal introduced at A) the remaining test point signals should be sine waves at the same frequency. There will

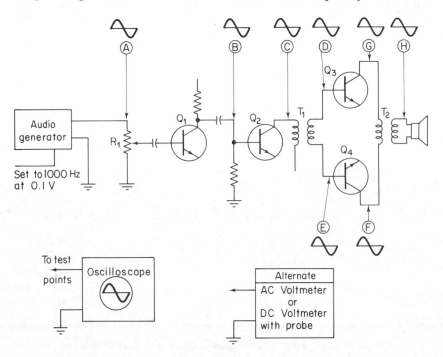

**Figure 4-27:** Basic test connections for troubleshooting audio/modulator section

## Sec. 4-3 Audio/Modulation Section Troubleshooting

probably be considerable voltage gain at points B and C, but you can only guess how much.

You can monitor each of the test points with an ac voltmeter, a dc voltmeter with rectifier probe, or a scope. We shall use the scope because it will show any really abnormal distortion at each of the test points, as well as any gain.

Armed with this information, you are now ready to begin the troubleshooting effort. You may take notes as you go along if that will help. However, keep in mind that each troubleshooting problem is always slightly different from the last. There is no surefire step-by-step procedure that will fit every situation.

*Determine the symptoms.* A symptom of no modulation (but good RF output) and no audio from the speaker is reported to you, and you then check the set yourself (never trust anyone). You find that the symptom of no modulation during transmit is correct, and the symptom of no audio output during receive is correct when the volume control $R_1$ is set at its midrange position (which is the normal operating position for $R_1$). However, by rotating $R_1$ fully on (maximum volume), you note that there is a very weak tone from the speaker. Thus, the no-audio during receive symptom is not absolutely correct, but it is obvious that the amplifier is not operating properly. (With $R_1$ at maximum and a 0.1 V input, the output should be over 3 W, and the tone would probably rattle your eardrums.)

Reset $R_1$ to its normal midrange position, and make your first decision.

*You could decide that the generator is bad and that there is no input signal at test point A.* This is not logical. When the volume control is turned to maximum volume, there is a weak tone from the speaker. Thus, although this output is weak, it does tell you that there is an input signal present. If there were no tone whatsoever, you could say that the audio generator might be bad or that you had made the connections improperly. Just to satisfy yourself, connect the scope across the input terminals (test point A), and observe the input waveform. It should be a sine wave at a frequency of 1,000 Hz, with an amplitude of about 0.1 V.

*You could decide that the power supply is defective.* This is slightly more logical than the decision concerning the generator. If the power supply is bad, you will get a symptom of no output tone whatsoever when $R_1$ is set to maximum volume. Because there is a weak output, *possibly* resulting from stray coupling around the defective stage or from a weak transistor, the power supply is working. Of course, the power supply could be producing a low voltage to one or more stages. To confirm this, you must check the voltage at each of the stages. However, at this point in the troubleshooting, remember that you are trying to determine the symptoms, not isolate the trouble.

*You should decide to check the output of the circuit group.* This is the first step in isolating the trouble to a circuit.

*Isolate the trouble to a circuit.* This is done by checking the waveform at the output (test point H) of the circuit group. The test connections are shown in Fig. 4-28. If the amplifier is operating properly, there should be a sine wave at H with an amplitude of about 3.1 V when $R_1$ is set to midrange. If the sine wave is present but there is no tone in the speaker, the speaker is suspect. If a replacement speaker is available, check by substitution. If no speaker is available, connect a 3.2 Ω resistor (as shown in Fig. 4-28) across the secondary terminals of $T_2$ (to substitute as the load), and observe the sine wave at H. If the sine wave is correct with a substitute load, the speaker is at fault. *Do not operate* the audio/modulation section without a load (either the speaker or the resistor). To do so could damage the transistors (particularly $Q_3$ and $Q_4$).

If the sine wave is absent at H with $R_1$ set to midrange, you can place a bad-output bracket at test point H (as discussed in Chapter 1). We

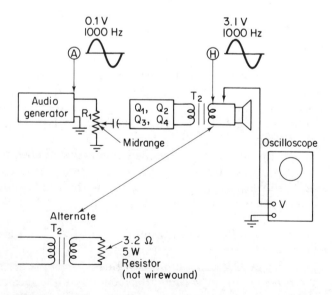

**Figure 4-28:** Test connections used to isolate trouble to circuit group (audio/modulator section)

Sec. 4-3  Audio/Modulation Section Troubleshooting     237

have a good-input bracket at A and a bad-output bracket at H; now it is time to make another decision.

*You could decide that the output transformer $T_2$ is defective.* This is possible but not logical. First of all, the trouble must be isolated to a single circuit by checking the input and output points. Thus far, you know there is a normal input at test point A (the input to the circuit group) and an abnormal output at test point H (the output of the circuit group). The trouble is located somewhere between these points, somewhere within the four circuits preceding test point H, *possibly* even in $T_2$. If the trouble is $T_2$, and you find it immediately, it would be a lucky guess, not logical troubleshooting. Further testing must be done before you can say definitely that any one circuit is defective.

*You could decide to use the half-split technique and make the next check at test points D or E.* Assume that you monitored point D and found no signal present (with $R_1$ at midrange). This is definitely an abnormal condition, but it proves very little. The problem could be associated with the circuits of $Q_1$, $Q_2$, or $Q_3$. Even the $Q_4$ circuit is not definitely eliminated (until you check at E). The fact that by chance you have selected a test point yielding an abnormal signal does not make your procedure correct.

*You could decide to use the half-split technique and make the next check at test point B.* This is a fairly logical choice. You must make a test at B sometime during the troubleshooting sequence. If the signal at B is abnormal, you have isolated the trouble to the $Q_1$ circuit (in one lucky jump). However, if the signal is normal at B, you are left with three possible defective circuits ($Q_2$, $Q_3$, or $Q_4$). There is a more logical choice.

*You should decide to use the half-split technique and make the next check at test point C.* This is the most logical choice because you have isolated the trouble to one-half of the audio/modulation section (or two circuits) in one jump. If the signal at C is abnormal, the trouble is in $Q_1$ or $Q_2$. If the signal at C is normal, the trouble is isolated to $Q_3$ or $Q_4$. (Note that the primary winding of $T_1$ is considered part of the $Q_2$ circuit, but that the secondary winding of $T_1$ is part of the $Q_3$–$Q_4$ circuit.)

Keep in mind that the terms *normal* and *abnormal* applied to the signals at test points B to G are arbitrary and relative. The service literature does not tell you the signal amplitudes or the correct waveforms. However, it is reasonable to assume that all the signals are sine waves (at least ac voltages at the frequency of the signal introduced in test point A). It is also reasonable to assume that there is some voltage gain at each test point as you proceed along the signal path. With an input of 0.1 V, the service literature says you can expect an output of 3.1 V. This is a voltage gain of about 30. Most of the gain (at least half, probably more) is obtained in the $Q_1$ and $Q_2$ circuits (test points B and C), because the $Q_3$ and $Q_4$ circuits are essentially power amplifiers. Similarly, the circuit of $Q_1$ should show more voltage gain than the circuit of $Q_2$, because the emitter resistor of $Q_1$ is bypassed by $C_2$. Thus, the *difference in gain* should be greater between points A and B than between points B and C. However, because exact values are not available, you must ultimately isolate the trouble with voltage-resistance checks, component checks, and the like.

*Assume that you have made the check at test point C and find the signal abnormal.* If you are paying attention, your next logical step is to monitor the signal at test point B. There is no reason to monitor any other test point under these conditions. The signals in the paths beyond point C will be abnormal if point C is abnormal.

*Now, assume that you have made the check at test point B and find the signal normal.* You have now isolated the trouble to a circuit (the $Q_2$ circuit), and you have done so in three logical jumps (from test point H to C to B). Your next step is to locate the specific trouble in the $Q_2$ circuit.

*Locate the specific trouble.* As discussed in Chapter 1, the first step in locating the specific trouble after the circuit has been isolated is to perform an inspection using the senses. As well as can be determined, the transistor is good because there is no evidence of physical damage (examine to see if this is true), and there is no evidence of overheating (touch the transistor; it should *not* be hot). There is no indication of burning components (no characteristic burning smell), and there are no obvious physical defects. You can conclude, therefore, that the inspection using the senses points to no outward sign of where the trouble is located. You must then rely on test procedures to locate the defective part.

Figure 4-29 shows both the physical relationship of the $Q_2$ circuit parts and the point-to-point wiring. This illustration is similar to that provided in well-prepared CB service literature. Count yourself lucky if you have such data when troubleshooting all CB sets.

It is now time to make another decision concerning your next step in troubleshooting.

*You could observe the waveform at the emitter of $Q_2$.* This would be of little value. With a normal amplifier circuit, the emitter waveform is

Sec. 4-3 Audio/Modulation Section Troubleshooting 239

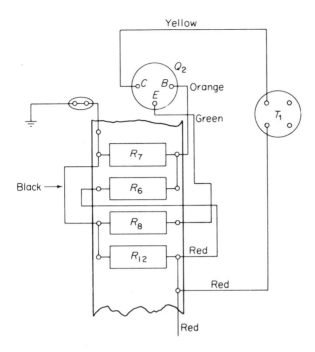

**Figure 4-29:** Physical relationship of $Q_2$ circuit parts and point-to-point wiring (practical wiring diagram)

similar to the collector waveform (test point C), except that the emitter usually shows lower amplitude. Because there is a low-amplitude waveform (or no amplitude) at the collector, you will find nothing of value at the emitter.

*You could decide that transformer $T_1$ is defective.* To prove your assumption, you could measure the voltage at the collector of $Q_2$. The schematic (Fig. 4-26) shows that the voltage should be 12 V. If there is no voltage at the collector of $Q_2$ or the voltage is very low, $T_1$ is probably defective (the primary is probably open or shorted). Your assumption is logical, but you are ahead of yourself. What if the voltage at the collector of $Q_2$ is correct? Then, $T_1$ is probably good (at least the primary is not open or shorted).

The sequence you have just performed should have taught you one thing: Do not make hasty decisions with regard to faulty parts. Make your decisions after you have gained enough information from the proper tests and measurements. There is a more logical choice than faulting transformer $T_1$ immediately.

You should check the voltages at each element of $Q_2$. Figure 4-30 shows the test connections for measuring the voltages. Note that the negative (−) terminal of the voltmeter is connected to a ground terminal on the chassis and that the positive (+) terminal is connected to each of the transistor elements in turn. This is so because the power supply is positive with respect to ground, as indicated on the schematic.

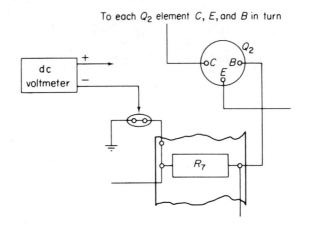

**Figure 4-30:** Test connections for measuring voltages at $Q_2$ elements

Figure 4-26 shows that the voltages should be +8, +8.7, and +12 V, respectively, for the emitter, base, and collector. If all the voltages are normal or nearly so, it is logical to assume that transistor $Q_2$ is at fault. You can make an in-circuit test of the transistor as described in Sec. 4-1.4, or you can substitute a known-good transistor, whichever procedure is most practical. (If the transistor fails the in-circuit test, you must try substitution.)

*If the voltage at one or more of the $Q_2$ elements is abnormal, you must now make resistance checks.* As you know from the discussion in Chapter 1, this is done by measuring the resistance from each element of the transistor to ground, using the resistance charts supplied in the service literature as a reference. But assume that you have no resistance charts; furthermore, the schematic does not give enough information for you to calculate the resistance-to-ground for each $Q_2$ element. A possible exception is the emitter. Here, the resistance should be 100 Ω because only $R_8$ (a 100 Ω resistor) is connected to the emitter. Both the base and the collector have several resistances in parallel. In any event, arriving at the correct resistance-to-ground value will only be through a wild guess.

## Sec. 4-3 Audio/Modulation Section Troubleshooting

Under these circumstances, continuity and resistance checks are your best bet. Let us examine each element of $Q_2$ in turn.

The collector of $Q_2$ is connected to the power supply through the primary winding of $T_1$. To check continuity in this line, disconnect the collector lead and make the connections shown in Fig. 4-31. Remember that you do not know the resistance of the $T_1$ primary winding, but you should have a continuity indication, *probably* in the order of a few ohms. If the ohmmeter shows an infinite resistance (with the range selector set on one of the high-resistance scales), the $T_1$ primary winding is probably open. If the ohmmeter resistance is zero (on the lowest scale), the $T_1$ primary winding is probably shorted. Keep in mind that you can skip this resistance measurement if the collector of $Q_2$ shows a normal voltage (about +12 V).

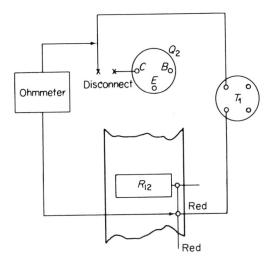

**Figure 4-31:** Test connections for checking continuity of $Q_2$ collector circuit

The emitter of $Q_2$ is connected to ground through $R_8$. To check continuity here, disconnect the emitter lead and make the connections shown in Fig. 4-32. The resistance should be about 100 Ω. Keep in mind that because all resistors have some tolerance, the reading will probably never be exactly 100 Ω. Again, a high or infinite reading indicates an open, whereas a low or zero reading indicates a short. From a practical standpoint, resistors usually do not short, but they do open.

The base of $Q_2$ is connected to the power supply through $R_6$. To check continuity in this line, disconnect the lead and make the test connections shown in Fig. 4-33. This will remove any parallel resistance from $R_7$ or $Q_2$. The resistance should be 3.3 kΩ.

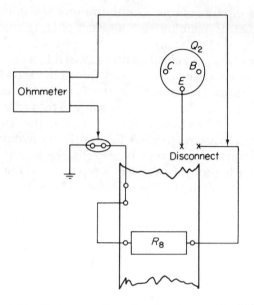

**Figure 4-32:** Test connections for checking continuity of $Q_2$ emitter circuit

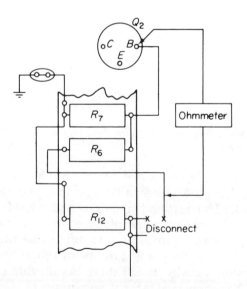

**Figure 4-33:** Test connections for checking continuity of $Q_2$ base circuit to power supply

Sec. 4-3   Audio/Modulation Section Troubleshooting    243

The base of $Q_2$ is connected to ground through $R_7$. To check continuity in this line, disconnect the lead and make the test connections shown in Fig. 4-34. This will remove any parallel resistance from $R_6$ or $Q_2$. The resistance should be 8.7 k$\Omega$.

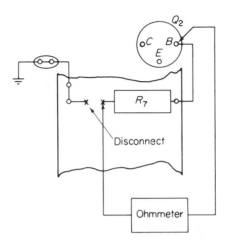

**Figure 4-34:**   Test connections for checking continuity of $Q_2$ base circuit to ground

Now, assume that the continuity-resistance checks show that $R_8$ is open. As discussed in Chapter 1, your next step is to make the necessary repairs and perform an operational check.

*Repairs and operational check.* After you have reviewed all the data and are satisfied that you have located the specific cause of the trouble, you should then repair the trouble. In this case, you should replace emitter resistor $R_8$ with a known-good resistor. What is your next step?

*You could turn on the power and make waveform and voltage measurements at $Q_2$.* Your first step after repairing the trouble should *not* be to turn on the power to the set. Rather, you should first verify that the repair you have made is good. That is, because this trouble was an open emitter circuit, you should check out the emitter circuit (continuity measurement) to be sure that the repair you have made is complete.

*You should, before turning on the power, measure the resistance of $R_8$ and check continuity from the emitter of $Q_2$ to ground.* Because this trouble was an open emitter circuit, you should first check out the emitter circuit for proper resistance and continuity. If by some strange chance the emitter resistance is still abnormal, you may assume that the repair you have made is not proper (or that you are incorrectly interpreting the resistance reading). Whatever the reason, you must take another look at your procedure. However, if the resistance and continuity from the emitter to ground are normal, you may be reasonably certain that you have properly repaired the trouble.

*Once proper resistance and continuity are established, you can turn on the power and make an operational check. What is the first thing you would do while making this check?*

*You could make voltage measurements at all elements of $Q_2$.* It is unnecessary to make these measurements at this time. You are now performing the operational check, not trying to isolate trouble to a defective branch of a circuit. You are fairly sure that the trouble is repaired; now, you simply want to verify this fact.

*You should check to be sure that all controls and switches, including those of the test equipment, are set for normal operation.* Always make sure that all switches and controls are first set for normal operation when you are performing an operational check. If the volume control ($R_1$ in Fig. 4-26) is set for minimum volume, there will be no sound from the speaker and no waveforms available at the test points, even though the amplifier is operating properly. Similarly, if the output control on the audio generator (the signal injection source) is accidentally set for minimum during troubleshooting, there will be no sound and no waveforms. Because those wrong control settings may be misinterpreted as trouble, it is important to set all controls for normal operation before attempting to perform the operational check.

## 4-4 RECEIVER SECTION TROUBLESHOOTING

Before presenting a step-by-step example of receiver troubleshooting, we shall discuss basic test and alignment procedures for the receiver section.

### 4-4.1 Oscillator Injection Signals to the Receiver

Keep in mind that the receiver must have injection signals from other sections of the CB set to operate properly. In the simplest receivers, such as those found in walkie-talkie sets (Sec. 3-6), there is a single local oscillator signal that is mixed with the incoming signal to produce the IF signal. In dual-conversion sets (and most CB sets are dual-conversion)

Sec. 4-4   Receiver Section Troubleshooting                                245

there are two oscillators (Sec. 3-1 through 3-5). In SSB receivers there is another oscillator that replaces the missing carrier (Sec. 3-5). As a first step in troubleshooting the receiver, check that all oscillator signals are available and are of the correct frequency and amplitude. The frequencies may be found from the schematic or other parts of the service literature. As discussed in Chapter 6, CB service literature usually lists the crystal and/or synthesizer frequencies. However, you may have to guess at the oscillator signal amplitudes. Typically, solid-state CB oscillator voltages are 1 V or less, and rarely above 1.5 V.

### 4-4.2   ANL Circuit Troubleshooting

Most CB receivers are provided with some form of automatic noise limiter. In most cases, the ANL is a simple series diode between the detector output and the audio section or volume control. In other sets, there is a separate noise signal channel that controls the audio. Either way, a failure in the ANL circuits can make it appear that the receiver section (RF, IF, and detector) is malfunctioning. For this reason, always check the ANL circuits before launching into the receiver section.

First, disable the ANL circuit and check operation. If this restores operation, the trouble is probably in the ANL circuit rather than in the receiver section. On some sets the ANL may be disabled by means of a switch. In other sets, it is necessary to short across the ANL series diode. If audio passes through with the ANL diode shorted, check the diode and associated parts by substitution.

### 4-4.3   Squelch Circuit Troubleshooting

Most CB receivers are provided with some form of squelch circuit. Although there are many types of squelch circuits, they all set the receiver input signal level at which the audio will pass. If the squelch is set too high, the receiver section will appear to be dead, or if the squelch circuits are defective, the receiver section may seem to be defective.

Before starting any receiver section troubleshooting, try resetting the squelch control. Usually, clockwise rotation of the squelch control silences the set, whereas counter-clockwise rotation will allow audio to pass. If the receiver appears to be dead with the squelch set full counter-clockwise, check out the squelch circuit.

*Basic squelch circuit test.* The following test may be applied to most sets equipped with a squelch circuit.

1. Connect a signal generator to the receiver input and set the generator output to zero (or tune to another channel).

2. With the volume at midrange, rotate the squelch control to the point that just silences the audio output completely.

3. Bring up the signal generator output slowly, and observe that the squelch breaks (audio passes through). Typically, if the squelch is set at the point where audio is just disabled, the squelch will open with a 1 $\mu$V (or less) signal at the receiver input.

4. If a large RF input signal is required to open the squelch (with the squelch properly set), or if the set remains dead with the squelch fully off, then it is practical to proceed with troubleshooting.

### 4-4.4 Receiver Section Alignment

The alignment procedures for CB set receivers are essentially the same as for any other AM receiver. If the set includes SSB, the set is operated in the AM mode during alignment, since alignment of an SSB receiver is complex (compared to AM). Complete alignment of the receiver section is usually not required; the RF and IF sections require only a "peaking" of the transformers and tuned circuits. Of course, there are exceptions to the rule. If there has been tinkering by untrained "technicians", or if you have replaced major parts (transistor, transformers, etc.) in the RF and IF sections, the sections may require extensive alignment. Keep in mind that replacement transformers (such as the 455 kHz IF transformers) are tuned to the correct frequency when they are shipped from the factory.

Always follow the service literature instructions when aligning the receiver section. In the absence of any instructions, the following procedures may be applied to most CB receivers, including those with PLL and SSB circuits.

*Oscillator circuits.* Always set (or check) the oscillator circuits before attempting alignment. This applies to all oscillator circuits that provide injection signals to the receiver (including both local oscillators for dual-conversion sets and the carrier reinsertion oscillator for SSB). The oscillators should be at peak output, and must be providing the correct signal frequencies.

*AGC or AVC circuits.* Most CB set receivers have some form of AVC or AGC circuit which may interfere with alignment. The purpose of an AVC/AGC circuit is to increase receiver sensitivity when weak input signals are applied, and vice versa. Thus, when you tune a transformer for a peak output reading, the AVC/AGC circuit will tend to flatten the reading. In some cases, it may be necessary to disable the AVC/AGC circuits by applying a "clamp" or bias voltage to the AVC/AGC line, as is done in TV servicing. As a rule, the AVC/AGC line may be clamped by

## Sec. 4-4 Receiver Section Troubleshooting

applying a dc voltage of the same polarity as the line, with a value of about twice the normal AVC/AGC line voltage. For example, if the AVC/AGC line is at +2 V, clamp the line with a +4 V signal.

Do not apply a clamp or bias to the AVC/AGC line if you get good peak readings when the RF and IF circuits are adjusted. If you must clamp the line, do not apply excessive voltages. Start by applying the normal or typical AVC/AGC voltage, and then increase as necessary to disable the AVC/AGC function, as shown in Fig. 4-35.

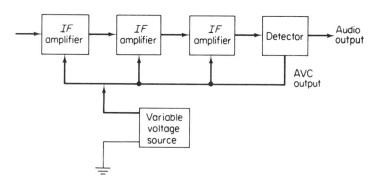

**Figure 4-35:** Connections for disabling the AVC line

Be very careful in choosing the *bias reference point*. Usually, it is preferable to use the emitter of the transistor as a reference, rather than chassis ground. In both examples of Fig. 4-36, chassis ground cannot be used as a reference, since even a low or zero bias would place a substantial difference of potential across the base-emitter junction the instant the variable voltage source test lead is touched to the transistor base.

*Output indicator for alignment.* The receiver S-meter may often be used as the output indicator for alignment procedures. In most sets, the S-meter is at or after the last IF stage. Unfortunately, the S-meter may also be in the RF stages. The next most popular arrangement is to connect an ac voltmeter across the loudspeaker voice coil. It is necessary that the alignment signals be modulated with an audio tone. Of course, if the service literature recommends a specific point for monitoring the output during alignment, always follow the literature.

*Basic receiver alignment procedure.* It may be necessary to modify the following procedure somewhat to match the circuits of all CB set receivers.

1. Set *all* of the oscillators that feed injection signals to the receiver. Oscillators should be at correct frequency and peak amplitude.

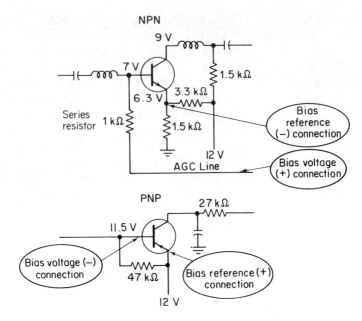

**Figure 4-36:** Proper bias connections for transistor stages

2. Connect a signal generator to the receiver antenna input and set the generator output to the correct channel frequency.

3. Modulate the generator output with a 1 kHz signal at 30 to 40% modulation.

4. Turn the squelch off and/or increase the signal generator output amplitude until a good midscale reading is obtained on the output indicator.

5. Adjust the windings of the IF transformers (capacitor or tuning slug) in turn, starting with the last stage and working toward the first. Adjust each winding for maximum reading. Repeat the procedure to make sure that there is no interaction between adjustments (usually there is some interaction, so you must compromise). Also, the service literature for some sets calls for monitoring an output point that produces a minimum (or null) indication when the circuits are properly tuned.

6. Repeat the procedures of step 5 for the RF transformers or circuits.

7. If you cannot get a good output indication with a signal applied to the antenna (due to severe misalignment or some other malfunction), set the generator output to the frequency of the IF stages (typically 455 kHz). Apply the generator signal to the input of the last IF stage, and adjust that stage for maximum. Then move to the next-to-last stage and tune for maximum. Continue until all of the IF stages are properly tuned. Repeat the procedure to check for interaction.

8. With the IF stages properly adjusted, repeat the procedure of step 7 for the RF transformers or circuits (using the correct frequencies).

Sec. 4-4   Receiver Section Troubleshooting   249

### 4-4.5   Example of Receiver Section Troubleshooting

This step-by-step troubleshooting problem involves locating the defective part in a solid-state receiver section. The problem has been localized to the receiver section using the basic procedures of Chapter 1 and Fig. 1-7. The trouble symptoms are: "no reception on any channel, transmission normal;" or, "reception is poor and transmission is good."

*General instructions.* The schematic diagram of the receiver is shown in Fig. 4-37. Note that an IC is used for the IF stages. Thus, if a fault is traced to any circuit within the IC, it is necessary to replace the entire IC unit. The test points shown on Fig. 4-37 do not appear in the service literature schematic, but they have been assigned arbitrarily to illustrate the troubleshooting procedure.

When test points are not shown on the schematic (which is typical for CB schematics), it is good practice to assign test points as a guide for troubleshooting. In general, use the collector and bases of transistors (or plates and grids of vacuum tubes) when test points are not indicated on the schematic. It is entirely possible to troubleshoot the receiver section (and most other sections of the set) using only this minimum information. Consider the following.

You know that the input RF signal is at the selected CB channel frequency. The schematic shows that the IF is at 455 kHz. From a troubleshooting standpoint, this establishes certain conditions.

You know that an RF signal may be introduced at test point A. This signal must be at the CB channel frequency. The same is true for test point B. However, if you inject a signal at B, it must be much greater in amplitude because $Q_1$ acts as an RF amplifier and normally supplies considerable voltage gain. If you are monitoring the signal at B, it should be identical to that at A except for amplitude.

The signal at test point D, the local oscillator injection signal, is at the CB channel frequency, plus or minus 455 kHz. The schematic indicates that the local oscillator injection frequency is at the CB channel frequency, minus 455 kHz. Since the receiver is single-conversion, only one oscillator injection signal need be considered.

Signals at test points C and E will be at 455 kHz. From a monitoring standpoint, the signal at E should be much larger in amplitude than that at C because of the gain produced in the IC amplifier stages. The signal at test point C may also include the local oscillator frequency developed by $Q_2$.

The signals at test points F and G are in the audio range. Thus, from the standpoint of signal injection, you must inject an AF signal at F.

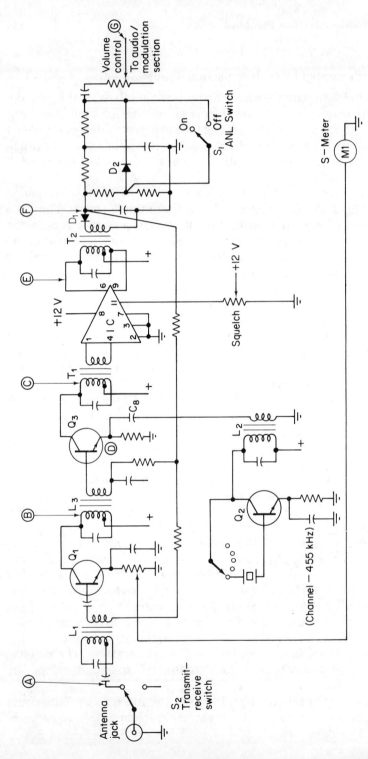

**Figure 4-37:** Schematic diagram with test points of receiver section

## Sec. 4-4 Receiver Section Troubleshooting

There is little point in injecting a signal at G, since this is the input to the audio/modulation section, and the trouble symptom stated that transmission was normal. As a possible exception, the output from G (the volume control) passes through contacts of a switch or relay to the audio/modulation section input. It may be that the contacts are bad.

If test points F and G are used for signal tracing, you must use an ac or a dc meter with a probe (or a scope) to monitor the signals. Also, if the incoming RF signal (either a CB station signal or a signal from an RF generator connected to some point ahead of E) is unmodulated, there will be no signal at E, even though the receiver is operating properly. Thus, if an RF generator is connected to A and you expect to monitor the signal at E with a scope or an ac meter, you must modulate the RF signal generator output.

With this limited information scrounged from the schematic (and your vast knowledge of CB receivers), you are now ready to start the troubleshooting.

*Determine the symptoms.* The user reports that no signals can be heard on any channel. You confirm this symptom, but you then check further.

The schematic shows that the ANL circuit can be cut out when the ANL switch $S_1$ is set to OFF (switch closed to short $D_2$). You find that with the ANL switch OFF, there is still no improvement.

You rotate the squelch control full off and recheck for signals at known active channels. There is still no improvement.

You check the S-meter indication on all channels, or on known active channels. If the S-meter is working in the normal manner (the S-meter indication varies as channels are changed, and it is moving over the scale), it is reasonable to assume that all stages up to the S-meter are good (at least they are operating). An S-meter check is of real value when it is operated from one of the last stages in the receiver (such as the last IF stage), since you can clear several circuits at once if you get a good S-meter indication. Unfortunately, in our example, the S-meter monitors the first stage $Q_1$. A good S-meter indication proves only that $Q_1$ is probably operating properly.

You rotate the volume control full on and notice that some noise can be heard at the loudspeaker. You then put your finger on test point G (the volume control arm) and notice that there is considerable hum in the loudspeaker. What is the most logical test point at which to begin troubleshooting? To simplify this example, you can eliminate test points A and G. In effect, you just checked G when you touched the volume control arm. The loudspeaker hum, coupled with the original trouble symptom of "good transmission" clears the audio/modulation section and the wiring between G and the audio input (through relay or switch contacts, etc.). Of course, if you insist, you can inject an audio signal at G (about 15 or 20 mV should be sufficient), and positively clear the audio/modulation section. Test point A is eliminated as a first test point to check, since A is for signal injection only.

You could choose test point F. If you monitor at F, and get a good signal, this points to a bad ANL circuit. However, this same condition would probably show up when you set the ANL switch to OFF. If you get no signal at F, it proves very little. The problem is still somewhere between A and F. If you inject a signal at F (it must be audio, probably at 1 V or less, at 1 kHz) and it passes, this proves only that the ANL circuits are probably good. However, it is possible that a shorted or badly leaking ANL diode $D_2$ will pass audio, so the test is not conclusive.

You could choose test point D. This is not a bad choice, if you intend to align the receiver section. As discussed in Sec. 4-4.4, one of the first steps in alignment is to monitor the local oscillator injection signal. However, as a first choice for troubleshooting, a good signal at D proves very little (only that the local oscillator signal is at the correct frequency and is of the correct amplitude). A bad signal at D would pinpoint the problem immediately, but this would be a lucky guess.

You could choose test point B. If you inject a signal at B and the response is good (that is, if you inject a modulated RF signal and hear the tone on the loudspeaker), you have isolated the trouble to the RF voltage amplifier $Q_1$, and you are also very lucky. (The good S-meter indication will tell you just about as much.) On the other hand, if the response is bad with a signal injected at B, you still have many circuits to check. You could also monitor the signal at B, but this requires two test instruments (a signal generator at A and a monitoring meter or scope at B).

You could choose test point E. If you inject a signal at E and the response is bad (that is, if you inject a modulated RF signal and do not hear the tone on the loudspeaker), you have isolated the trouble to the detector $D_1$ circuit, and you are again very lucky. If the response is good with a signal injected at E, you still have many circuits to check. Again, you could monitor the signal at E, but this also requires two test instruments.

There is another problem with test point E. You choose to inject an audio signal at E. If $D_1$ and the associated circuit parts are good, the audio signal should pass as if it were injected at F. However, it is possible that a shorted or badly leaking $D_1$ will also pass the straight audio signal, but will not pass the audio portion of the modulated RF signal.

Sec. 4-5  Transmitter Section Troubleshooting    253

*You should choose test point C.* By choosing C, after you have checked the ANL, squelch, S-meter, and volume control, you have effectively split the receiver section in half. You are on the right track. But do you inject a signal at C, or monitor the signal at C? Many technicians will argue either way. But because injection is simpler (in this case) you choose to inject a 455 kHz modulated signal at C, and you find that there is a good response (you hear a tone on the loudspeaker and there is plenty of volume).

The trouble is now isolated to three circuits, the RF voltage amplifier $Q_1$, the local oscillator $Q_2$, or the mixer $Q_3$. To isolate the trouble further, there are two obvious (and equally logical) steps. You can inject at B, or monitor at D. You choose to inject at B, using a modulated RF signal at the channel frequency. If the response is good, you have isolated the trouble to the RF voltage amplifier $Q_1$. But you are not lucky this time, and the response is bad (there is no tone on the loudspeaker).

Your next step is to monitor at D. Now you are lucky, and there is no local oscillator injection signal at D. You quickly conclude that $Q_2$ is bad. You are wrong. $C_8$ is open, which you locate by moving the monitoring device to both sides of $C_8$. The signal is present and normal at the oscillator side of $C_8$, but not at D. You replace $C_8$ and restore normal operation to the receiver section.

## 4-5  TRANSMITTER SECTION TROUBLESHOOTING

Before presenting a step-by-step example of transmitter section troubleshooting, we shall discuss basic test and alignment procedures for the transmitter section.

### 4-5.1  Basic RF Tests

The following paragraphs describe test procedures for RF amplifiers used in the transmitter section. The first part is devoted to test and measurement procedures for the resonant circuits used at radio frequencies (resonant-frequency measurements, Q measurements, etc.). The remaining sections cover test procedures for complete RF amplifiers used in the transmitter section.

For best results, RF amplifiers are tested with all components soldered in place, both before and after troubleshooting. This will indicate whether there is any change in circuit characteristics resulting from the physical relocation of components during troubleshooting or repair procedures.

Often, there is capacitance between components, from components to wiring, and between wires. These stray components may add to the reactance and impedance of circuit components. When the physical locations of parts and wiring are changed, the stray reactances change and alter circuit performance.

### 4-5.2 Basic RF Voltage Measurement

As discussed in Chapter 2, when the voltages to be measured are at radio frequencies and are beyond the frequency capabilities of a meter or scope, an RF probe is required (Sec. 2-5). Such probes rectify the RF signals into a dc output that is approximately equal to the RF voltage. The dc output of the probe is then applied to the meter or scope and is displayed as a voltage readout in the normal manner.

If a probe is available as an accessory for a particular meter or scope, use that probe rather then any homemade device. The manufacturer's probe is matched to the meter or scope in calibration, frequency compensation, and the like. If a probe is not available for a particular meter, use the probe described in Sec. 2-5.9 and Fig. 2-16.

*If the frequency of the voltage is to be measured,* it is possible to do so using an oscilloscope. However, it is far more practical to use a frequency counter such as that described in Sec. 2-6.

### 4-5.3 Measuring LC Circuit Resonant Frequency

The circuit for measuring a resonant frequency of an LC circuit is shown in Fig. 4-38.

To use the circuit, adjust the unmodulated RF generator output amplitude for a convenient indication on the meter. Then, starting at a frequency well below the lowest possible frequency of the LC circuit, slowly increase the generator output frequency.

For a parallel-resonant LC circuit, watch the meter for a maximum or peak indication.

For a series-resonant LC circuit, watch the meter for a minimum or dip indication.

The resonant frequency of the LC circuit is the one at which there is a maximum (for parallel) or minimum (for series) indication on the meter.

Note that there may be peak or dip indications at harmonics of the resonant frequency. The test is most efficient when the approximate resonant frequency is known.

To broaden the response (so that the peak or dip may be approached more slowly), increase the value of $R_L$ from 100 k$\Omega$. (An increase in $R_L$ lowers the LC circuit Q). To sharpen the response, lower the value of $R_L$.

Sec. 4-5   Transmitter Section Troubleshooting                               255

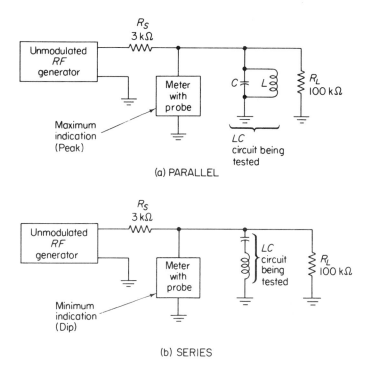

Figure 4-38: Measuring resonant frequency of LC circuits

### 4-5.4 Measuring Inductance of a Coil

The circuit for measuring inductance of a coil is shown in Fig. 4-39. To use the circuit, adjust the unmodulated RF generator output amplitude for a convenient indication on the meter. Then, starting at a frequency well below the lowest possible resonant frequency of the LC combination being tested, slowly increase the generator output frequency.

The resonant frequency of the LC circuit is the one at which there is a maximum indication on the meter. Using this resonant frequency and a known capacitance value, calculate the unknown inductance using the equations given in Fig. 4-39. To simplify calculations, use a convenient capacitance value, such as 100 pF or 1,000 pF.

Note that when a known inductance value is available, the procedure can be reversed to find an unknown capacitance value.

Increase the value of $R_L$ to broaden the peak indication if desired, or decrease $R_L$ to sharpen the peak.

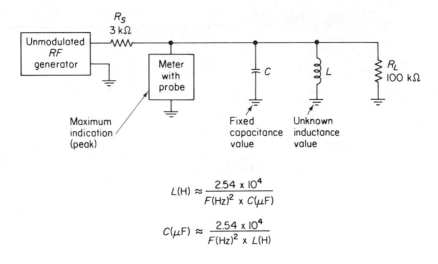

$$L(H) \approx \frac{2.54 \times 10^4}{F(Hz)^2 \times C(\mu F)}$$

$$C(\mu F) \approx \frac{2.54 \times 10^4}{F(Hz)^2 \times L(H)}$$

**Figure 4-39:** Measuring inductance of a coil

### 4-5.5 Measuring Self-resonance and Distributed Capacitance of a Coil

There is distributed capacitance in any coil, which may combine with the coil's inductance to form a resonant circuit. Although the self-resonant frequency may be high in relation to the operating frequency at which the coil is used, it may be near a harmonic of that operating frequency. This limits the usefulness of the coil in an LC circuit. Some coils, particularly RF chokes used in transmitters, may have more than one self-resonant frequency.

The circuit for measuring self-resonance and distributed capacitance of a coil is shown in Fig. 4-40. To use the circuit, adjust the unmodulated RF generator output amplitude for a convenient indication on the meter. Tune the generator over its entire frequency range, starting at the lowest frequency. Watch for either peak or dip indications on the meter. A peak or dip indicates that the inductance is at a self-resonant point. The generator output frequency at which this point occurs is the self-resonant frequency (or a harmonic).

Make certain that peak or dip indications are not the result of changes in generator output level. Cover the entire frequency range of the generator, or at least from the lowest frequency up to the third harmonic of the highest frequency involved in circuit design or operation.

Once the resonant frequency (or frequencies) has been found, calculate the distributed capacitance, using the equations given in Fig. 4-40.

Sec. 4-5 Transmitter Section Troubleshooting 257

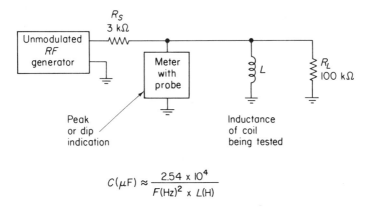

$$C(\mu F) \approx \frac{2.54 \times 10^4}{F(Hz)^2 \times L(H)}$$

**Figure 4-40:** Measuring self-resonance and distributed capacitance of a coil

### 4-5.6 Measuring Q of Resonant Circuits

The Q of a resonant circuit sets the circuit band widths. That is, a high Q circuit has a narrow band width (sharp tuning), whereas a low Q circuit has a wide band width (broad tuning). The tuning circuits of RF amplifiers used in CB transmitters are generally high Q. Possible exceptions are the tuning circuits of RF linear amplifiers used in SSB transmitters.

From a practical standpoint, it is not necessary to measure the Q or bandwidth of transmitter circuits when troubleshooting CB sets. However, you should know the procedure. Also, an understanding of the procedure may help you to locate some obscure defect in transmitter circuits when all the usual tests (recommended in the service literature) appear normal.

The most practical measurement of resonant circuit Q is the bandwidth at the resonant frequency. The circuits for measuring bandwidth (or Q) of resonant circuits are shown in Fig. 4-41. Figure 4-41a shows the test circuit in which the signal generator is connected directly to the input of a complete stage; Fig. 4-41b shows the indirect method of connecting the signal generator to the input.

When the stage or circuit has sufficient gain to provide a good reading on the meter with a nominal output from the generator, the indirect method (with isolating resistor) is preferred. Any signal generator has some output impedance (typically 50 Ω). When this resistance is connected directly to the tuned circuit, the Q is lowered, and the response becomes broader. (In some cases, the generator output impedance can seriously detune the circuit.)

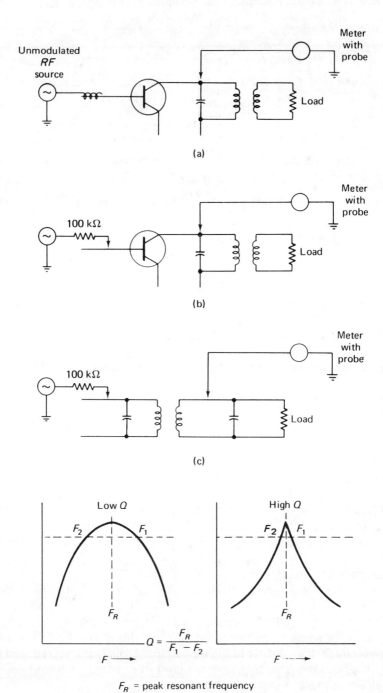

$F_R$ = peak resonant frequency

**Figure 4-41:** Measuring Q of resonant circuits

Sec. 4-5 Transmitter Section Troubleshooting 259

Figure 4-41c shows the test circuit for a single component (such as an RF transformer). When the resonant circuit is normally used with a load, the most realistic Q measurement is made with the circuit terminated in that load value. A fixed resistance may be used to simulate the load. The Q of a resonant circuit is often dependent upon the load value.

To use the circuit, adjust the unmodulated RF generator output amplitude for a convenient indication on the meter. Tune the signal generator to the approximate resonant frequency of the circuit; then, tune the generator for maximum or peak reading on the meter. Note the generator frequency.

Tune the generator below resonance until the meter reading is 0.707 of the maximum reading. Note the generator frequency (frequency $F_2$). To make the calculation more convenient, adjust the generator output level so that the meter reading is some even value (such as 1 V or 10 V) after the generator is tuned for maximum. This will make it easy to find the 0.707 mark.

Tune the generator above resonance until the meter reading is 0.707 of the maximum reading. Note the generator frequency (frequency $F_1$).

Calculate the circuit Q using the equation given in Fig. 4-41.

### 4-5.7 Testing Transmitter Section RF Amplifier Circuits

There are as many ways to test the circuits of the transmitter section as there are circuit variations; however, certain basic requirements must be met.

*Dummy load or RF wattmeter.* During normal operation, the final output stage of the transmitter section is connected to the antenna. During troubleshooting, it is often convenient to disconnect the antenna so that undesired signals are not broadcast. When the antenna is disconnected, it *must be* replaced by a dummy load (Sec. 2-7) or an RF wattmeter (Sec. 2-8). The RF wattmeter has an advantage over a dummy load, since the wattmeter contains a load and provides an indication of the output power from the final stage. This knowledge can be quite helpful in troubleshooting. For example, if you key the transmitter and find a 4 W output indication on the RF wattmeter, you know that the transmitter section is operating. Of course, the transmitter may be off frequency, the modulation may be absent or incorrect, etc. Also, if you find no indication, or a low-power indication, when the transmitter is keyed, you know that the transmitter section is malfunctioning. A dummy load, by itself, will not provide this type of information.

*Meter or scope with RF probe.* If an RF probe is not available, or as an alternative, it is possible to use a test circuit such as the one shown in Fig. 4-42. This circuit is essentially a pickup coil (which is placed near the RF amplifier inductance) and a rectifier that converts the radio frequency into a dc voltage for measurement on a meter. One problem with the circuit in Fig. 4-42 is that some RF circuits are shielded or contained in a shielded can. This prevents the loop from picking up enough RF to give a good indication. Such circuits must be tested with an RF probe (and meter or scope).

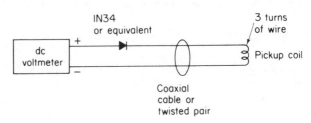

**Figure 4-42:** Circuit for pickup and measurement of RF signals (alternate to RF probe)

*Frequency measurement.* Although it is possible to measure transmitter signal frequencies with a scope, the procedure is somewhat elaborate and may be difficult, particularly at the high frequencies involved (27 MHz). To keep the process as simple as possible, use a frequency counter (Sec. 2-6) for all transmitter section frequency measurements.

*Basic RF transmitter section test procedures.* Figure 4-43 shows the basic circuit for test and measurement of RF amplifier circuits. The circuit also provides for testing the oscillator and for drive signals applied to the transmitter from frequency synthesizers and/or PLL circuits. If the transmitter section amplifiers being tested do not have an oscillator or a drive signal from another section, a drive signal must be supplied. An RF signal generator may be used to supply the drive signal, provided the generator can produce a signal of correct amplitude and frequency. Typically, drive signals to the transmitter section are in the 27 MHz range and about 1 V or less. However, there are exceptions.

Connect the probe or pickup to each amplifier stage in turn. Start with the first stage (this will be the oscillator for simple sets or the input from the frequency synthesizer in most other sets) and work toward the final or output amplifier stage.

A voltage indication should be obtained at each stage. Usually, the voltage indication increases with each amplifier stage, as you work toward the antenna. However, some stages may be frequency multipliers and therefore provide no voltage amplification.

If a particular amplifier stage is to be tuned, adjust the tuning control for a maximum reading on the meter or scope. It should be noted that this

Sec. 4-5  Transmitter Section Troubleshooting 261

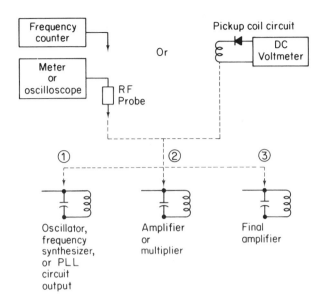

**Figure 4-43:** Connections for test of RF circuits during transmitter section troubleshooting

tuning method does not guarantee that each stage is at the desired operating frequency. The method does show that a signal is present and that the circuit is tuned for peak. However, it is possible to get maximum readings on harmonics. Fortunately, RF transmitter amplifier circuits are usually designed so that they will not tune to both the desired operating frequency and a harmonic. Generally, RF amplifier tank circuits tune on either side of the desired frequency, but not to a harmonic (unless the circuit is seriously detuned).

After each circuit is tuned for peak, check the signal frequency at each stage with a frequency counter, starting with the oscillator or frequency synthesizer and working toward the antenna. Keep in mind that the final amplifier output signal to the antenna must be at the desired channel frequency, within 0.005%.

Once you are satisfied that the output frequency is within tolerance on all channels, check the power output with an RF wattmeter. The final amplifier output signal to the antenna must be no more than 4 W.

*Using the S/RF meter for transmitter test and adjustment.* Obviously, the S/RF meter on the CB set front panel provides a quick check of transmitter section operation. If you get a good RF indication when the transmitter is keyed (on each channel), it is reasonable to assume that the transmitter is operating. However, you must still check the output signal to the antenna for correct RF power (with an RF wattmeter) and frequency (with a frequency counter).

In some cases, it is possible to adjust the transmitter section using the S/RF meter. You simply key the transmitter and adjust each transmitter circuit for a peak reading; however there are some drawbacks to this procedure. First, you must check that the S/RF meter monitors the final output to the antenna and that there are no adjustment controls after the monitoring point; you can check these points on the schematic. Even if the S/RF meter monitors the final output, you must check the RF power output and frequency after adjustment. Also, if any of the transmitter section circuits are seriously detuned, it may be difficult to get a good reading on the S/RF meter, and you must use the procedures shown in Fig. 4-43 for each stage.

### 4-5.8 Modulation Measurement

In addition to checking the transmitter section for correct frequency and power output, the percentage of modulation should also be checked. This test should be made even though the audio/modulation section performs properly during receive. There are several reasons for this; for example, the audio/modulation section may produce low gain that would get by during receive (should the volume control be set toward maximum). The audio/modulation section output transformer may have two windings (or two sets of connections), one for transmit and one for receive. (The receive winding might be good and the transmit winding bad.) The microphone might be defective or weak.

It is important to know that most CB sets contain some circuits (with adjustment controls) that set the percentage of modulation. These modulation control circuits are often external to the audio/modulation section, since they are not used during receive. In some cases, the modulation control circuits are a form of clipper that limits input to the audio/modulation section during transmit. In other cases, the modulation is controlled by feedback from the audio/modulation section output to input. Either way, a failure in the modulation control circuits may make the audio/modulation section to appear defective.

*Direct or trapezoidal modulation measurement.* An oscilloscope may be used to measure the percentage of modulation. The two basic methods, direct and trapezoidal, are discussed in Sec. 2-3. In either method, remember that the modulation *must not* exceed 100% in any case. Typically, the modulation is above 85% when normal voice is used directly into the microphone. Overmodulation or undermodulation may be corrected using the methods described below.

Sec. 4-5   Transmitter Section Troubleshooting                          263

*Overmodulation.* Generally, overmodulation is caused by improper adjustment and may be corrected by proper adjustment. Of course, it is possible that the modulation control circuits have failed (no clipping, no feedback, etc.), but this is less likely than improper adjustment. Always check the modulation control circuits first when overmodulation cannot be corrected by adjustment.

*Undermodulation.* When the audio/modulation section appears normal during receive, but there is undermodulation during transmit, the most likely causes are: (1) improper adjustment, (2) a defective microphone (or not speaking directly into the microphone), or (3) defects in the modulation control circuits, in that order. Thus, you should always try to correct undermodulation by adjustment or by a check of the microphone, before launching into the control circuits.

*Microphone checks.* The simplest check of a microphone is by substitution. When this procedure is not practical, try measuring the output voltage produced by the microphone for normal voice signals. Typically, speaking directly into high-impedance microphone (such as the ceramic type) will produce about 15 to 30 mV on a 1 M$\Omega$ audio-range voltmeter. Under the same test conditions, it should be possible to obtain 150 to 300 mV by whistling loudly into a good high-impedance microphone.

### 4-5.9   Example of Transmitter Section Troubleshooting

This step-by-step troubleshooting problem involves locating the defective part in a solid-state transmitter section. The problem has been localized to the transmitter section using the basic procedures outlined in Chapter 1 and Fig. 1-7. The trouble symptoms are "no transmission on any channel, reception normal"; or, "transmission is poor, reception good."

*General instructions.* The schematic diagram of the transmitter section is shown in Fig. 4-44. The test points shown do not appear in the service literature schematic, but have been arbitrarily assigned to illustrate the troubleshooting process. As discussed for receiver section troubleshooting (Sec. 4-4.5), it is entirely possible to troubleshoot the transmitter section using only the minimum information found on the schematic diagram (and this may be all you get).

No voltage or resistance information (at transistor terminals) is available except for the +12 V supply voltage found on the schematic. However, you should be able to calculate the voltages found at the

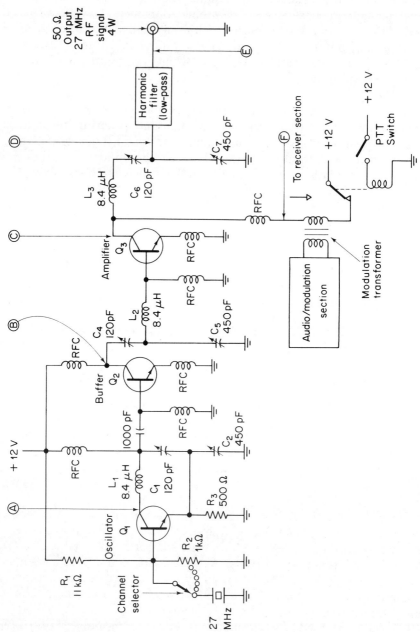

**Figure 4-4:** Schematic diagram with test points of transmitter section

## Sec. 4-5 Transmitter Section Troubleshooting

transistor terminals. The collectors of the three transmitter section transistors ($Q_1$, $Q_2$, and $Q_3$) are all connected to the +12 V through RF chokes. Such chokes generally have very little dc resistance and thus produce very little voltage drop (usually a fraction of a volt drop). Thus, it is reasonable to assume that the dc voltage at the collectors is about +12 V (or slightly less).

The emitters of $Q_2$ and $Q_3$ are connected to ground through RF chokes. Thus, the dc voltage of the emitters should be zero, and the resistance to ground should be a few ohms at most. Possibly some dc voltage might be developed across the chokes, but it is not likely. The same is true of the $Q_2$ and $Q_3$ bases.

The base of $Q_1$ has a fixed dc voltage applied through the voltage-divider network of $R_1$ and $R_2$. The ratio of $R_1$ to $R_2$ indicates that the voltage drop across $R_2$ is about 1 V. Thus, the base of $Q_1$ is about 1 V. Typically, the emitter of $Q_1$ will be within 0.5 V of the base (assuming that $Q_1$ is a silicon transistor). The resistance to ground from the emitter of $Q_1$ should be equal to $R_3$, or about 500 $\Omega$.

The schematic shows that the output is 4 W into a 50 $\Omega$ antenna and the crystals are in the 27 MHz range, so that all three stages are tuned to the channel frequency, and there is no frequency multiplication in any stage.

Capacitors $C_1$, $C_2$, and $C_4$ through $C_7$ are tuning adjustments. (Normally, each capacitor is adjusted for a peak output reading.) There are no other operating or adjustment controls. Once you press the microphone PTT switch, the transmitter section should perform its function and transmit AM-modulated signals to the antenna.

With this wealth of information to draw upon, you are ready to plunge into the troubleshooting by determining the symptoms.

*Determine the symptoms.* In some troubleshooting work, you will find that the set was never properly tuned, adjusted, or otherwise correctly put into operation. Some troubleshooters approach each new problem with this assumption. However, in most CB problems, you will find that the set had been working satisfactorily for a time before the trouble occurred. This is the case in our example; although the set had been operating properly for many months, the operator reports to you that there is now no output. The transmission cannot be heard by any CB stations tuned to the same channel. What is your first step?

*You could start by checking each circuit to determine which one is not operating.* This procedure rates a definite no, even though you would eventually locate the faulty circuit. It is not a logical approach and would probably require several unnecessary tests before the trouble was isolated to a circuit.

You could perform the tuning procedure to determine which circuit group does not perform properly. This would be even more illogical than checking each stage. Although you might locate the trouble area using this procedure, it would probably require many unnecessary steps. Each step should provide the most information with the least amount of testing.

*You should check the output.* Notice that we do not tell you how to check the output at this time, but you have two methods for doing so: with or without test equipment. You may also check the output with the antenna connected or disconnected.

The first place to look is the S/RF meter on the CB set front panel. A good RF output indicates that the transmitter section is operating. The power may be low and the signal may be off-frequency, but there is RF output. Unfortunately, as you can see in Fig. 4-44, there is no S/RF meter in our example.

With the antenna still connected, the output may be checked by tuning a CB receiver to the same channel, keying the suspected CB transmitter, and trying to transmit a signal (voice into microphone). Although this test is quick and easy, it does not provide all the information you need.

Even if you hear the voice transmission on the test CB receiver, it is possible that the transmitter output is low. A weak transmission might be picked up by a nearby receiver but could not be heard over the normal communications range. If you do not hear the transmission on the receiver, it proves that the operator was right, but you have no clue as to which section of the set is at fault. The problem could be in the audio/modulation section.

By using a field strength meter (Sec. 2-10) or one of the special test sets described in Sec. 2-14 that includes a field strength function, you can leave the antenna connected and check the output signal.

It is possible that the antenna or lead-in is defective. To verify this, check the output with a dummy load and meter or scope (including an RF probe), or with an RF wattmeter. The wattmeter is best for checking the RF output, since it will provide an immediate answer to the question "is there RF output and is the power correct?" With the dummy load and meter or scope, you must measure the voltage and calculate the power. Typically, you will get about 14 V of RF voltage with a 50 $\Omega$ antenna or load. Power is equal to $E^2/R$. Thus, $14^2/50 \approx 4$ W.

Even if you do get the correct power output, it is still possible that the transmitter section is not being modulated by the audio/modulation section. Thus, the most positive test of the transmitted output is to

## Sec. 4-5 Transmitter Section Troubleshooting

monitor the output on a scope, using one of the percentage of modulation test methods described in Sec. 2-3. The most accurate results are obtained by modulating the set with an audio tone. However, for our example you are more interested in seeing if the set will perform its normal function of transmitting voice signals; you should get between 85 and 100% modulation with normal voice.

Now, assume that you disconnect the antenna, connect a dummy load, and check percentage of modulation using one of the methods described in Sec. 2-3. There is no RF indication on the scope (no vertical deflection whatsoever) with the transmitter keyed (PTT button pressed) and someone speaking directly into the microphone. In which section of the set do you think the trouble is located?

*Modulation section.* If you choose the audio/modulator you will probably not succeed as a troubleshooter. You could try being an investment consultant or opening a health food store. Or you are not paying attention. Or you simply do not understand transmitters. The vertical deflection on the scope is produced by the RF signal; the shape of the signal is determined by the modulation. Thus, with no vertical deflection, there is no RF signal, and the trouble can be traced to the transmitter section.

*Transmitter section.* You have made a logical choice. The next step is to isolate the trouble to one of the circuits (oscillator $Q_1$, buffer $Q_2$, or power amplifier $Q_3$ in the transmitter section). What is the next most logical test point?

*You could check the supply voltage for the transmitter section.* This approach does have some merit. Note that $Q_1$ and $Q_2$ receive their collector voltage directly from the +12 V supply line, whereas $Q_3$ gets its collector voltage through a secondary winding on the modulation transformer and through relay contacts (which are operated by the PTT switch). Thus, to make a complete check of supply voltage for the RF circuits, you must measure the voltage at each of the three collectors. Of course, if the voltage at any one collector is present and correct, the power supply itself is good.

*You could check for RF signals at test point A.* This is not a bad choice. If there is no RF signal at A, the trouble is traced to the oscillator ($Q_1$ and associated parts). If there is RF at A, you know that the oscillator is operating, but you eliminate only one circuit as a possible trouble area. Thus, it is not the most logical choice.

You could check for RF signals at test points C, D, or E. This would prove very little, since, in effect, you are checking at these points when you monitor the output signal at the antenna.

You should check for RF signals at test point B. This is the most logical choice, since it is essentially a "half-split" of the transmitter section (Chapter 1). As shown in Fig. 4-45, you can check with a meter or scope and an RF probe. If the scope is capable of passing the 27 MHz signals, you can use the alternate test method shown in Fig. 4-45b. However, the probe is usually the most practical method.

If there is a good RF signal indication at B, both $Q_1$ and $Q_2$ are operating properly, and the trouble is traced to the power amplifier $Q_3$, or possibly the low-pass TV interference filter. You also eliminate the

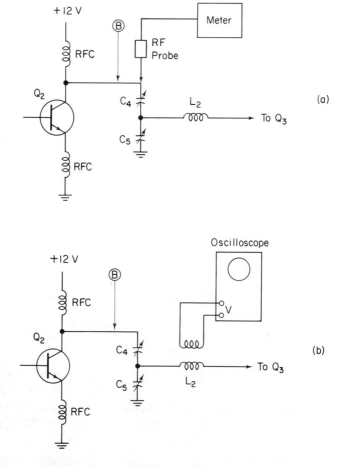

**Figure 4-45:** Checking for RF signal at test point B

## Sec. 4-5 Transmitter Section Troubleshooting

power supply, because it must be good if $Q_1$ and $Q_2$ are functioning normally. However, you do not eliminate the line between the power supply and the collector of $Q_3$ (which is a separate path from that between the supply and the $Q_1$–$Q_2$ collectors).

If there is no RF indication at B, the trouble is traced to $Q_1$ or $Q_2$. The next step is to check for RF signals at A.

Now, assume that there is a good RF indication at B. The power amplifier ($Q_3$ and associated circuit parts) may be bracketed as the faulty circuit because the input signal is good and the output is bad. To confirm this, check for RF signals at C, D, and E, on the off chance that there is a good signal at C and D, but not at E (due to a bad low-pass filter).

*Locate the specific fault.* Now that the trouble is isolated to the circuit, it must be located. First, perform a visual inspection of the power amplifier. Note that $Q_3$ is probably provided with a heat sink or is mounted on a chassis that acts as a heat sink. Assume that there is no apparent sign of where the trouble is located. There is no sign of overheating, and all components (as well as wiring) appear normal. What is your next step?

*You could make an in-circuit test of $Q_3$.* This would be difficult because $Q_3$ is operated class C and will not respond to the usual in-circuit forward-bias tests (as discussed in Sec. 4-1.4). No forward bias is applied to $Q_3$, so you cannot remove the bias. If you attempt to apply bias, the voltage relationship between emitter and collector will probably not change because there is no dc load. An in-circuit transistor tester might prove that the transistor is good (at audio frequencies). However, in-circuit testers are generally useless at radio frequencies.

*You could make a substitution test of $Q_3$.* This would be more satisfactory than any in-circuit test. And it is possible that you may have to substitute $Q_3$ before you locate the fault. However, more convenient tests may be made at this time.

*You could check the resistance at all elements of $Q_3$.* Although it will probably be necessary to check resistance and continuity before you are through, resistance checks at this time will prove little. The resistance-to-ground at the emitter $Q_3$ should be a few ohms at most. The base resistance-to-ground should also be near zero. Only a high resistance-to-ground at the base and emitter would be significant.

*You should check the voltage at all elements of $Q_3$ first.* The voltage at the base and emitter should be zero (we are speaking of dc voltage, not RF signal voltage). The dc voltage at the $Q_3$ collector should be about

+12 V. If the voltages are all good, you may skip the resistance-to-ground measurements. However, you may still have to make continuity checks if the voltages are abnormal. Let us examine possible faults indicated by abnormal voltages.

*Large dc voltages at base or emitter.* Large dc voltages at the base or emitter of $Q_3$ indicate that the elements are not making proper contact with ground. For example, a high-resistance solder joint between the $Q_3$ emitter and ground indicates the possibility that a dc voltage may appear at the emitter. Also, if the emitter-ground connection is completely broken, the emitter will be "floating" and show a dc voltage.

*No dc voltage at collector.* If the collector shows no dc voltage, the fault is probably in the RF choke, the modulation transformer's secondary winding, or the relay contacts, indicating the need for a continuity check. Check the dc voltage at test point F. If the voltage is correct at F but not at C (the collector of $Q_3$), the RF choke is at fault. If the voltage is absent at F, the fault is in the transformer winding or the relay contacts.

Now, to summarize this troubleshooting example, let us assume that the trouble is caused by an open $L_3$ coil winding. This will not affect dc voltage or resistance. Substitution of $Q_3$ will not cure the problem. These are the kinds of problems you will find in real CB troubleshooting: Everything appears to be good, but the set will not work!

To solve such problems, you must make point-to-point continuity checks. In this case, if you checked from point C to the top of $C_6$, you would have found the open coil winding.

# 5
# SERVICING ANTENNA AND NOISE PROBLEMS

Many CB service problems result from external causes. The two major problem areas are defective antennas (or antenna lead-ins) and electrical noise (particularly noise or interference caused by automobile engines). Because of their importance, this entire chapter is devoted to these subjects.

## 5-1 ANTENNA SERVICE PROBLEMS

At least half the CB sets that show chronic "poor performance" symptoms are being used with defective antennas or have defective antenna installations. We will not go into the step-by-step details of installing a CB antenna (either base station or mobile). There are at least a dozen books and hundreds of magazine articles on the subject (including several by this author). Instead, we will concentrate on quick-check service methods to determine if the antenna is functioning properly and how to pinpoint the problem in a defective antenna system.

A CB antenna may not be performing properly because of a defect such as shorts in the coaxial lead-in, poor contact, open contact, broken shielding, etc., or it may have never performed properly because of a mismatch or improper adjustment. Either way, you will want to know if the antenna is capable of performing properly now.

### 5-1.1 SWR Meter Antenna Checks

The SWR meters described in Sec. 2-11 are the most useful test instruments for checking antennas. You can tell at a glance if the CB set, lead-in, and antenna are properly matched. If the CB set has a built-in SWR meter, use it. If not, use an external SWR meter or one of the special test sets (Sec. 2-14) that include an SWR function.

Ideally, the SWR should be checked on all channels. As a minimum, check SWR at the highest, lowest, and middle channels. An SWR reading of 1.1 or 1.2 on all channels indicates that the antenna system is good. Leave it alone! If the SWR readings are about 1.2 to 1.5, the antenna system is on the borderline.

*Adjustable antennas.* If the SWR reading is good on the lower channels but gets progressively worse (a higher SWR reading) on the higher channels, the antenna is probably too long. If the opposite is true (good SWR on the higher channels but not on the lower) the antenna is probably too short. Some, but not all CB antennas are adjustable. On an adjustable antenna, loosen an allen screw and shorten or lengthen the antenna as necessary. Also, the loading coils of some antennas are adjustable.

No matter how the antenna is to be adjusted, operate the set on the channel that shows the poorest SWR reading, and make the adjustment to get a good SWR reading on that channel. Then recheck SWR at the other channels. For example, if the high channels show poor SWR, operate on the highest channel (23 or 40) and adjust the antenna until the SWR is good (about 1.2 or 1.3, if possible). Then recheck at channel 1 for a good SWR reading. Adjust the antenna for the best possible SWR on all channels. This usually involves some compromise.

*Antenna matching devices.* A number of antenna matching devices are on the market. These devices are connected between the set and the antenna. In effect, the matching devices change the antenna length electrically. The author has no recommendations on any of the devices. In general, from a service standpoint, it is better to try correcting the antenna problem, rather than to compensate. Of course, if the antenna is good (mismatch at a minimum, proper continuity, no shorts, etc.) but not adjustable, and SWR is poor, an external antenna matching device may be the answer.

*RF amplifier tuning.* If the antenna is not adjustable, it is sometimes possible to detune (very slightly) the final RF amplifier of the transmitter

Sec. 5-1  Antenna Service Problems                                    273

section to get a better SWR. The success of this technique depends on amplifier tuning circuit design, considerable experimentation, and some luck. The usual sequence for tuning is to adjust the final amplifier for peak with an RF wattmeter and/or the set's S/RF meter and then check SWR. If there is an SWR problem, try detuning the final RF amplifier circuit and see if the SWR improves. Do not make any drastic changes in final amplifier tuning (from the peak). If drastic changes are required, a mismatch or other problems are present.

Always make sure the output frequency is within tolerance (0.005% of the channel frequency) and power output has not been seriously reduced, if the final amplifier is detuned in any way. Verify that the improvement in SWR holds for all channels, high and low. Finally, keep in mind that the technique will not work on all sets. Also, an antenna problem of major proportions cannot be corrected by adjustment of the CB set, nor should any attempt be made to do so.

### 5-1.2  Field Strength Meter Antenna Checks

The field strength (FS) meters described in Sec. 2-10 may also be used to test an antenna system. However, FS meters do not provide the proof available from an SWR meter. An FS meter shows the *relative* field strength of the transmitted signal. If the FS meter indication is drastically different for each channel, the antenna system is suspect. For example, if the FS meter indication is much stronger at higher (or lower) channels, check antenna adjustment as described in Sec. 5-1.1.

Keep in mind that the FS meter will not produce exactly the same indication on all channels, even in a perfect antenna system. Also, an FS meter indication that is the same for all channels does not prove that the antenna system is good. A severe mismatch in the antenna system may still produce a strong indication on a relative FS meter. Thus, an FS meter indication proves only that the antenna is radiating a signal.

### 5-1.3  Ohmmeter Antenna Checks

In most cases, a mobile CB antenna (automobile or shipboard) may be checked with an ohmmeter by using a pair of extra long leads as shown in Fig. 5-1. The procedure is as follows:

*Antenna continuity.* Connect one ohmmeter lead to the antenna tip and the other to the inner conductor of the lead-in that plugs into the CB set. Use alligator clips to make the connection. Set the ohmmeter to the lowest scale, and then shake the antenna and lead-in. The resistance indicated by the ohmmeter should be about 3 to 5 ohms for a typical

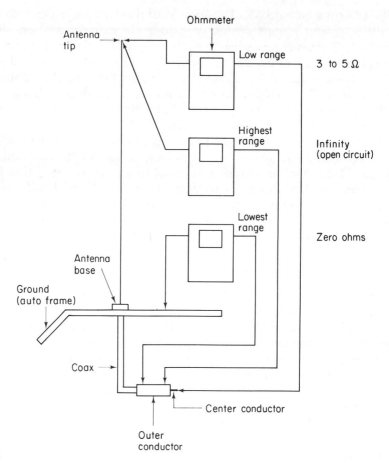

**Figure 5-1:** Checking mobile antenna installations for continuity and shorts

mobile CB installation. The resistance may be slightly higher for some installations.

If the resistance is considerably higher, or if it varies when either the antenna or lead-in is moved, there is poor contact between the antenna and lead-in. This reduces antenna efficiency and may result in chronic "poor performance" of the CB set, both during reception and transmission. An antenna connection that is intermittently making and breaking contact will produce a popping or cracking noise in the loudspeaker when the car (or boat) is in motion.

If you suspect high resistance or poor contact, disconnect the lead-in

Sec. 5-1  Antenna Service Problems    275

from the antenna and measure the lead-in resistance and the antenna resistance separately. Shake, bend, or twist the individual units to see which one is causing the problem. Look for corrosion or dirt at the coaxial connectors of the lead-in. Lead-ins and their connectors are the most likely source of poor contact. However, a collapsible antenna may have poor contact between segments, and an antenna loading coil may make poor contact with the antenna or base.

*Antenna shorts and grounds.* Once you are satisfied that there is proper continuity from the antenna to the lead-in center conductor, make certain *there is no contact* between the antenna and the outer conductor or shield. Connect one ohmmeter lead to the antenna tip and the other to a ground or the lead-in outer conductor, as shown in Fig. 5-1. Set the ohmmeter to the highest range and shake the antenna and lead-in. The ohmmeter should read infinity (open circuit).

If an intermittent short is indicated, a visual check will usually pinpoint the source. The problem is not so easy to locate when the ohmmeter indicates a constant high resistance that does not change when the antenna and lead-in are moved. Such a high resistance short will reduce antenna efficiency during transmission and reception both and may be a noise source. Look for moisture at the connectors and for frayed braiding on the outer conductor of the lead-in. High resistance shorts may also be caused by an accumulation of dirt at the antenna base, since the dirt can hold enough moisture to form a high resistance between antenna and ground.

*Outer conductor continuity.* The outer conductor of a coaxial cable is both a conductor and a shield. A shield braiding that is broken or making poor contact, has the same effect as inner conductor with a bad contact (reduced antenna efficiency, noise, etc.). Connect one ohmmeter lead to the outer conductor of the lead-in at the CB set end and the opposite lead to ground. Set the ohmmeter to the lowest range and shake or bend the lead-in. The resistance should be zero or near zero. If not, the outer conductor is not making proper contact with the ground and is not performing its normal function of shielding the lead-in. In addition to reducing antenna efficiency, poor contact in the lead-in outer conductor offers an easy path for electrical noise to enter the set. This is a particular problem if the lead-in is routed near an engine or other electrical wiring.

If the outer conductor shows poor contact (a high resistance to ground), look for breaks at the point where the shielding enters the connectors. In most cases, a visual check will pinpoint the poor connections.

## 5-2 NOISE AND INTERFERENCE PROBLEMS

Electrical interference, or noise, particularly interference caused by automobile engines, has long been one of the most annoying factors in CB operation. CB manufacturers have taken what steps they can to solve this interference problem by adding shielding and filters to the sensitive portions of the receiver section. As discussed in Chapter 3, most CB sets include some form of squelch, ANL, and other noise-suppressing circuits. Effective as these circuits are, they only set the noise level and make it bearable. Weak signals that are below this level simply do not get through. The result is a limiting of the operating range of the receiver section and a possible loss of communication.

It is impractical to discuss all causes and cures of electrical interference in one chapter; an entire book would be necessary. This author's book *Eliminating Engine Interference*, 2nd ed., published by Howard W. Sams and Co., Inc., Indianapolis, Indiana, has long been the standard in this field, and no attempt to duplicate this work will be made here. Instead, we will summarize interference problems and remedies from a service standpoint.

### 5-2.1 Causes of Engine Interference

All engine electrical interference or noise originates from the same source—sparking. The gasoline engine is full of spark-noise sources. Spark plugs, distributor points, distributor contacts, alternator slip rings, generator brushes, and voltage-regulator contacts all have gaps that electricity must jump in the normal operation of an engine. Electric sparks from any source are the chief creators of noise interference, since sparks produce undesirable radio waves that can be radiated or conducted into the CB set. Obviously, sparking could be eliminated by stopping the engine or cutting off the alternator/generator each time the set is used, but this is not a practical solution.

The problem must be attacked at its source, the engine itself. Three basic methods may be used: arc suppression, filtering, and shielding. A number of noise eliminator kits are available, all based on these three techniques.

### 5-2.2 Identifying Interference Sources

It is possible to identify the source of interference by the sound that comes from the CB set loudspeaker. Here are the characteristics:

> Ignition interference (and this is usually about 90% of the problem) occurs as a popping sound that is synchronized with the speed of the engine.

Sec. 5-2 Noise and Interference Problems 277

- Alternator/generator noise (the next most common problem) may be identified as a "whine" that starts only when the engine is speeded up. If the noise source is doubtful, temporarily remove the alternator/generator leads (or slip the alternator/generator belt off), and check the noise levels with the engine running. If the noise is unchanged, the alternator/generator is not the source.
- A voltage regulator usually produces a rough, rasping sound as it cuts the alternator/generator in and out of the circuit.
- Instrument noise (caused by fuel gauges, etc.) is identified by hissing or crackling sounds. Instruments should be disconnected one at a time until the noise-maker is found.
- Tire and wheel noises usually result in an irregular "rushing" sound in the receiver. This sound is present only when the auto is in motion.

*Preliminary tests.* The easiest way to run a noise or interference test on any engine is to compare the receiver noise level while the engine is running with the noise level under identical conditions while the engine is off. Try to make the test away from an external noise source, such as other engines, high-voltage lines, neon lights, etc. Turn off the squelch and any other noise limiter circuits, if possible. Adjust the receiver volume control until you hear a steady noise level from the speaker. If there are any signals on the air, select a channel with a weak signal and adjust the volume until the signal is barely audible. Then turn off the engine and see if the noise level drops noticeably.

If there is no great change in background noise when the engine is turned off, or if there is no change in the weak signal, you do not have an engine interference problem. If you do note an appreciable change, start looking for the source of the interference. It is generally assumed that if you cannot hear any noise, there is no interference. This is not always true, since strong radiated interference can desensitize a receiver (particularly the IF amplifiers) so that weak signals are obscured. For that reason, make the noise tests on the basis of background noise level and a weak signal.

*Radiated or conducted interference.* Once you have established the presence of engine interference, your next step is to decide whether the problem is radiated or conducted. With the set operating and the volume set so that you can hear both background noise and signals, disconnect the antenna. If the noise is completely eliminated, the problem is probably one of radiated, rather than conducted, noise. On the other hand, if the noise remains but the signal is gone when the antenna is disconnected, you have conducted noise.

### 5-2.3 Arc Suppression at Spark Plugs and Distributor

Arc suppression is used to combat both conducted and radiated noise. Arc suppression systems or kits are based on three facts: (1) although it takes a high voltage to make a spark jump across the gap of a spark plug or distributor contact, very little current is needed; (2) the amount of interference is proportional to the current; and (3) a high resistance placed in series with the spark plug and distributor will limit the current to practically nothing without reducing the voltage.

In practice, arc suppression is accomplished by installing a suppressor at each spark plug. A suppressor is also added in the coil circuit at the distributor. Figure 5-2 shows a typical noise elimination system based

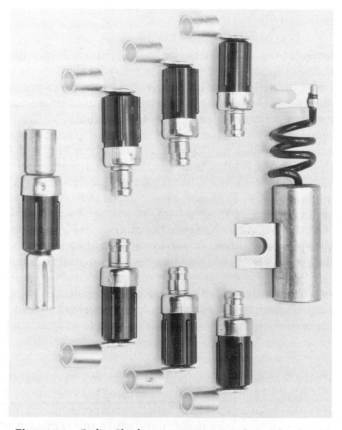

**Figure 5-2:** Radio Shack noise-suppression kit with plug-in suppressors and filter capacitor

Sec. 5-2  Noise and Interference Problems

on arc suppression. The right-angle suppressors are for the spark plugs, and the straight suppressor is for the distributor. The system or kit in Fig. 5-2 also includes a filter capacitor.

There are some drawbacks in using suppressors, even though they are effective. Most present-day autos use suppressor-type ignition wire and suppressor spark plugs. In some cases, the resistance of the external suppressors combined with that of the ignition wiring and plugs is sufficient to impair the engine. At the same time, the maximum resistance that can be added may not be enough to eliminate all objectionable interference. It is quite possible that filtering (for conducted noise) and shielding (for radiated noise) will be required in addition to (or instead of) arc suppressors.

### 5-2.4  Interference Filters for Ignition Systems

The three basic types of interference filters in common use today are filter capacitors, chokes, and wave traps.

The most popular type of *filter capacitor* is the feed-through type shown in Fig. 5-3. This device is essentially a bypass capacitor, but is

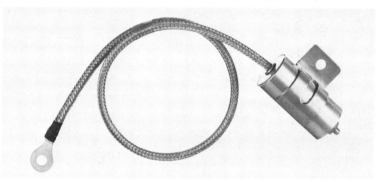

**Figure 5-3:**  Estes Engineering Company filter capacitor. (Estes Engineering Co., Port Angeles, Wash.)

described as a "shield dc power-line filter and cable" by the manufacturer. (Note that the noise elimination kit in Fig. 5-2 also includes a similar filter capacitor.) The filter in Fig. 5-3 is designed for use between the ignition system and the CB set. The feed-through bypass capacitor eliminates conducted noise in the battery circuit before the noise can enter and degrade or desensitize the receiver section of the CB set. The shielded cable attached to the capacitor eliminates noise pickup in the power cable from nearby spark plugs, alternators, generators, voltage regulators, etc.

Many noise elimination systems or kits include several bypass capacitors combined with suppressors in various points throughout the

**280**   Ch. 5   SERVICING ANTENNA AND NOISE PROBLEMS

ignition system. Figure 5-4 shows a typical example of suppressors combined with bypass capacitors and shielded wiring. Other systems include only bypass capacitors and shielded wiring, as shown in Figs. 5-5 through 5-7.

Some noise eliminator kits include *choke coils*. These kits are meant primarily for auto stereo units, rather than CB sets. The choke coil is connected in the dc power line between the set and the battery.

*Wave traps* are used primarily to filter alternator or generator interference, and they are more difficult to install than filters and suppressors (wave traps may require adjustment after installation, unless they are pretuned at the factory).

A wave trap consists of a coil and a capacitor combination (tank circuit) that is tuned to either the frequency of the spark interference or the operating frequency of the CB set. The wave trap rejects or traps interference of this frequency (similar to wave traps in antenna systems).

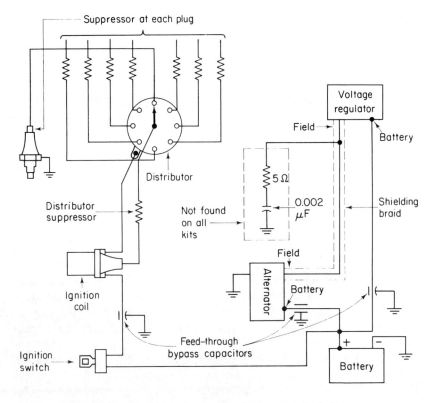

**Figure 5-4:** Typical noise elimination system including suppressors, bypass cpacitors, and shielded wiring

Sec. 5-2  Noise and Interference Problems     281

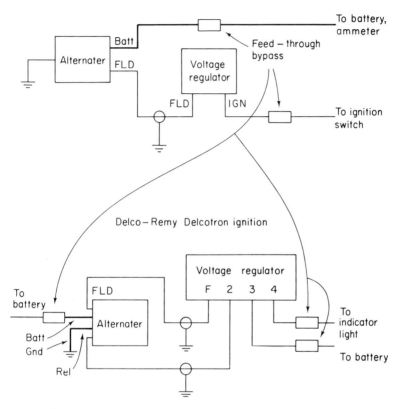

**Figure 5-5:** Estes Engineering Company noise elimination systems for Chrysler and Delco-Remy ignitions. (Estes Engineering Co., Port Angeles, Wash.)

The major drawback of this device is that it can be tuned to only one frequency, and spark noise is often generated on several frequencies at once. However, wave traps can be effective in the case of extreme alternator or generator interference on one specific channel (or a few adjacent channels).

Figure 5-8 shows installation of a typical wave trap on a generator or alternator. An alternator usually produces less interference than a generator and is less likely to require a wave trap. However, if alternator noise is present (as confirmed by disconnecting the alternator drive belt), an alternator noise filter such as shown in Fig. 5-9 can help to minimize noise on all channels. The alternator filter in Fig. 5-9 does not require tuning, and can thus be used with CB sets, stereos, etc.

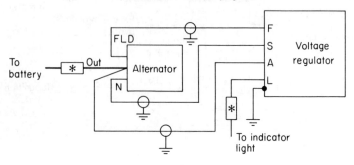

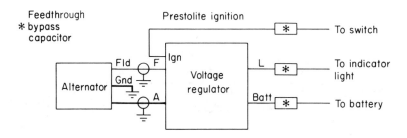

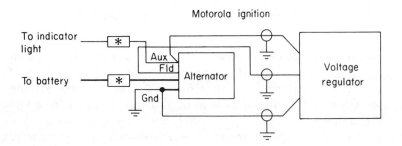

**Figure 5-6:** Estes Engineering Company noise elimination systems for Ford, Prestolite, and Motorola ignitions. (Estes Engineering Co., Port Angeles, Wash.)

Sec. 5-2 Noise and Interference Problems

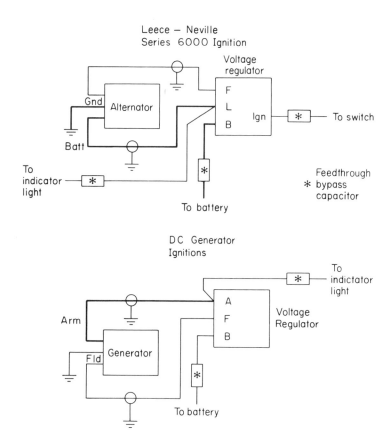

**Figure 5-7:** Estes Engineering Company noise elimination systems for Leece-Neville and DC generator ignitions. (Estes Engineering Co., Port Angeles, Wash.)

### 5-2.5 Shield Systems

Even when suppressors and filters are used, it is still possible for an engine to create sufficient interference to affect CB operation. Filtering is most effective on conducted noise (noise that is conducted from the ignition system to the set). Radiated noise is a greater problem.

As discussed, signals are generated when there is a spark across a gap. These signals are conducted through the lines in which the spark occurs, and they are radiated from the lines. Just as in the case of conventional radio waves, electrical interference may be prevented from radiating by surrounding the source of interference with a metal shield.

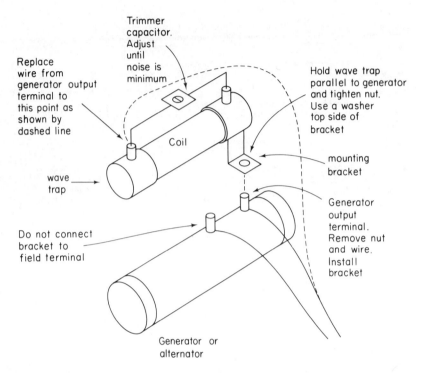

**Figure 5-8:** Installation of a typical wave trap on a generator or alternator

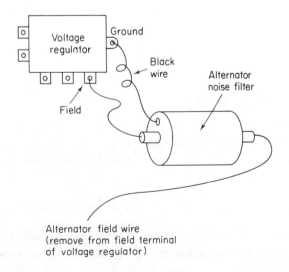

**Figure 5-9:** Installation of a typical alternator noise filter

Sec. 5-2  Noise and Interference Problems 285

There would be far more ignition interference on all mobile CB if it were not for the fact that the metal fire wall and auto body form a natural shield. The noise problem is more serious in marine engines because such shielding is lacking. Even in auto installations, however, the radiated interference is picked up by power wiring inside the engine compartment. From there, the noise is conducted through the fire wall to the CB set. The only sure way to prevent radiated interference is to surround the source and block the radiations. This is the reason for mechanical shielding systems in cases of extreme interference caused by the engine ignition system.

Shielding for typical mobile or marine installations is available in two types of kits—the fully assembled or "custom" kit, and the semiassembled kit.

The complete custom kits consist of metal shields for the spark plugs, high-voltage wiring with shielded braid, and shields for the distributor cap and ignition coil, all supplied as an assembled harness. Figure 5-10 is typical of the fully-assembled ignition shielding systems. Custom systems are higher in cost than the semiassembled systems, since it is necessary for the manufacturer to design parts and cut cables that will fit every make and model of engine.

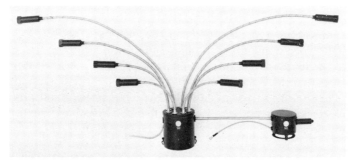

**Figure 5-10:** Estes Engineering Company fully-assembled ignition shielding system. (Estes Engineering Co., Port Angeles, Wash.)

A typical semiassembled shielding kit consists of spark-plug shields (each with several feet of shielding braid), a distributor-cap shield, and a shield for the ignition coil (with feed-through filter capacitors as necessary). These kits are sold with step-by-step instructions for installation, as are the custom kits. However, installation of the semiassembled kit is quite involved and time-consuming. For this reason, this type of kit is being replaced by the custom kit.

In addition to these two extremes in shielding kits, there are in-between installations, in which some of the components are assembled but other components must be fabricated on the spot. Also, many do-it-yourself systems are described in the various CB publications. The author cannot recommend any of the do-it-yourself systems.

### 5-2.6 Bonding Systems

Bonding provides an easy route for radiated interference to reach ground. The term *bonding*, as applied here, means connecting two metal objects with a metallic conductor so that there is a good electrical path between them. For example, both the hood and the frame of an auto are made of metal (except on some sports cars). This metal acts as a shield for containing interference radiated from the engine. However, if the hood is not connected electrically to the frame, the shield is not complete, and interference signals can be radiated through or around the hood. With proper bonding, a way is provided to route the interference from the hood to the frame, which acts as a ground.

There are two basic types of bonding: direct and strap. Direct bonding occurs when two metals are connected directly, surface to surface. Direct bonding is part of automobile design, and is not the problem of the service technician. However, strap bonding can be added to most autos as a noise elimination measure. Strap bonding is accomplished by connecting the two metals with an electrical conductor, usually a braid with lug fasteners at the ends. Braid is used because it is flexible and will not break easily with constant movement, and quite often the two metals to be bonded will move in relation to each other when the auto is in motion. This motion could break an ordinary solid electrical wire. As a side effect, the rubbing of two metals can produce static electricity, which creates interference under some conditions. When parts are bonded together, the static electricity has a discharge path through the strap.

*Typical bonding points.* Some typical bonding points on an automobile installation are:

> Corners of the engine to the frame
> The exhaust pipe to the frame and the engine
> Both sides of the hood
> Both sides of the trunk lid
> The coil and distributor to the engine and the fire wall
> The air cleaner to the engine block
> Battery ground to the frame

## Sec. 5-2 Noise and Interference Problems

The alternator or generator to the voltage-regulator frame

Front and rear bumpers to the frames on both sides

The tail pipe to the frame at the rear

### 5-2.7 Wheel and Tire Static

Both the wheels and tires of an auto are sources of static electricity. However, interference of this type will show up only when the auto is in motion. One way to pin down wheel or tire static is to switch off the engine and let the auto coast down a slight incline. Any interference still present (usually an irregular rushing sound) is probably from the wheels or tires.

In most cases, wheel static occurs in the front wheels and is caused by the insulating film produced by the lubricant in the wheel bearings. Commercial collector springs, installed in the hub cap, are available to offset this condition. The springs provide a grounding path from the rotating wheel to the axle.

There is also a static discharge between the auto tires and the road surface, particularly on hot, dry days. This may be eliminated by using antistatic powder. However, for a typical CB installation, such measures are not warranted.

### 5-2.8 Vibrator Interference

Vibrators are not required for modern solid-state mobile CB sets. However, most of the older vacuum-tube sets still use some form of vibrator. Since vibrators are mechanical devices that operate on the make and break of contacts, they can produce sparks and be a source of interference. One solution to the problem is to replace the mechanical vibrator with a transistor vibrator that contains no moving parts. Replacement solid-state vibrators come in a metal can identical in size and shape to the original mechanical vibrator, and are simply plugged into the same socket. One possible problem with solid-state vibrators is that they may not be able to handle the current of a mechanical unit. To be on the safe side, be sure to check the current rating of the replacement against that of the original.

### 5-2.9 Ignition System Maintenance

As a final consideration, remember that a poorly maintained ignition system will cause more interference than the ignition that is in good condition. For example, worn, ragged spark-plug gaps require higher ignition voltage, so both CB set and engine performance deteriorate.

Flat, parallel gaps and factory-maintained settings in spark plugs require less voltage and decrease interference. Careful spark-plug maintenance is therefore the first step toward ignition-noise suppression. The same applies to the distributor points, cap, rotor, ignition coil, alternator/generator, and voltage regulator. It is usually a waste of time and money for the service technician to cover up a poorly maintained ignition system with suppression, filtering, and shielding. Also, it is possible that when noise suppression is applied to a poorly maintained ignition, the engine will malfunction (rough idle, stall, etc.). Of course, a properly-installed noise elimination system should not affect a well-maintained ignition.

# 6

# CB SERVICE LITERATURE AND DATA

This chapter is devoted to understanding and using CB service literature. The need for service literature is obvious; however, good service data is not always readily available for all CB sets. For that reason, we shall describe service procedures for a cross section of CB sets. Using the data in this chapter, you can relate the procedures to the corresponding circuits of the set you are servicing.

For example, most CB sets have some means of adjusting the percentage of modulation (so as not to exceed 100%): we will describe such procedures. Although the exact details do not apply to all sets, there are similar modulation adjustment procedures that achieve the same results for any set. If you understand the typical procedures given here, you should have no trouble in making modulation adjustments for your set, even if the service literature is sketchy (or nonexistent).

## 6-1 CB CHANNEL FREQUENCIES

Figure 6-1 shows the frequencies for all 40 CB channels. Note that there is a minimum of 10 kHz between channels. Thus, if modulation is limited to 3 kHz (as it must to meet FCC regulations), there will be no overlap of sidebands between adjacent channels.

| Channel | Channel Frequency in MHz | Channel | Channel Frequency in MHz |
|---|---|---|---|
| 1 | 26.965 | 21 | 27.215 |
| 2 | 26.975 | 22 | 27.225 |
| 3 | 26.985 | 23 | 27.255 |
| 4 | 27.005 | 24 | 27.235 |
| 5 | 27.015 | 25 | 27.245 |
| 6 | 27.025 | 26 | 27.265 |
| 7 | 27.035 | 27 | 27.275 |
| 8 | 27.055 | 28 | 27.285 |
| 9 | 27.065 | 29 | 27.295 |
| 10 | 27.075 | 30 | 27.305 |
| 11 | 27.085 | 31 | 27.315 |
| 12 | 27.105 | 32 | 27.325 |
| 13 | 27.115 | 33 | 27.335 |
| 14 | 27.125 | 34 | 27.345 |
| 15 | 27.135 | 35 | 27.355 |
| 16 | 27.155 | 36 | 27.365 |
| 17 | 27.165 | 37 | 27.375 |
| 18 | 27.175 | 38 | 27.385 |
| 19 | 27.185 | 39 | 27.395 |
| 20 | 27.205 | 40 | 27.405 |

**Figure 6-1:** CB channel frequencies

## 6-2 INTERPRETING CB SERVICE LITERATURE

There is no standardization of service literature supplied by CB set manufacturers. At one extreme, the literature is in data-sheet form and includes (at best) a schematic diagram, a parts placement diagram, a parts list, a specification sheet, an operating procedure, and some installation data. At the other extreme, the literature is an instruction or service manual. Such manuals include all of the above data, plus a detailed theory of operation for all circuits, possibly some partial schematics, a crystal frequency chart showing which crystals are involved for each channel during both transmit and receive, complete alignment/adjustment instructions, voltage charts, resistance charts, test point alignment diagrams, injection voltages, and detector voltages.

An analysis of some sample alignment/adjustment procedures is given in Sec. 6-3. The discussion below describes other data found in typical service literature.

Sec. 6-2  Interpreting CB Service Literature            291

### 6-2.1  Schematic Diagrams

Figure 3-56 is typical of the full schematics found in well-done service literature. Note that the voltages are given for all elements of the transistors. In some service literature, voltages are given only in chart form. Always pay attention to any notes found on the schematics.

For example, two sets of voltages are given for transistor $TR_6$. One set is for "squelch on" and the other for "squelch off." It is common to give two or more sets of voltages, signals, or resistances when these values change for transmit and receive, or for other "on-off" conditions, such as squelch, ANL, etc.

### 6-2.2  Block Diagrams and Partial Schematics

The block diagrams and partial schematics given in Chapter 3 are typical of those found in some CB service literature. However, very few manufacturers provide both block diagrams and partial schematics; in fact, partial schematics are a rarity.

### 6-2.3  Parts Placement Diagrams

Figure 6-2 is typical of parts placement diagrams. Note that the parts are identified by reference designations and are superimposed on printed circuit (PC) board diagrams. Where the wiring is not printed circuit, some literature will include practical wiring diagrams such as shown in Fig. 4-29.

### 6-2.4  Parts Lists

Although there is no standardization, most parts are listed by reference designation, in alpha-numeric order ($C_1$, $C_2$, $R_1$, $R_2$, etc.). Thus, there is good cross reference to the parts placement diagrams such as shown in Fig. 6-2. The listing is followed by a description and a part number that is usually the CB set manufacturer's number, not the part number of the component manufacturer. For example, a part may be listed as: $C_7$, 16 pF ±5%, 1 kV, capacitor, fixed, ceramic, part number ABC-888.

Probably, the capacitor $C_7$ can be replaced by any 16 pF 1 kV fixed ceramic capacitor. However, if $C_7$ is used in any circuit of the transmitter section, or in any circuit that could affect the transmitted signal (its frequency accuracy, modulation percentage, power output, etc.), and the CB manufacturer's *exact replacement* is not used, it is possible that the set is no longer type-accepted and could be in violation of FCC rules.

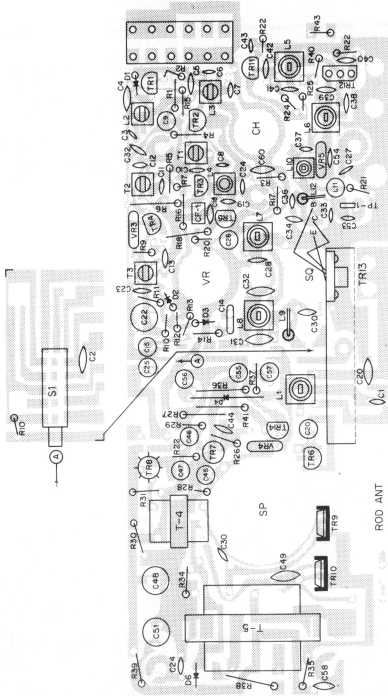

**Figure 6-2:** Typical parts placement diagram

## 6-2.5 Voltage and Resistance Charts

Figure 6-3 is typical of the voltage and resistance charts found in CB service literature. Note that there are two sets of voltages, one for transmit and one for receive. It is absolutely essential that all notes be ob-

### VOLTAGE CHART

| No. | Tube | ** | Pin 1 | Pin 2 | Pin 3 | Pin 4 | Pin 5 | Pin 6 | Pin 7 | Pin 8 | Pin 9 | Pin 10 | Pin 12 |
|---|---|---|---|---|---|---|---|---|---|---|---|---|---|
| V1 | 6DS4 | T | | | | | | | | | | | 6.3 VAC |
| | | R | | +100 | | −0.9 | | | 0 | | | 0 | 6.3 VAC |
| V12 | 12A27 | T | +280 | 0 | +3.8 | 0 | 0 | +80 | 0 | +5.3 | 6.3 Vac | | |
| | | R | +280 | 0 | +3.8 | 0 | 0 | +270 | +2.8 | +23.5 | 6.3 Vac | | |
| V13 | 12BH7 | T | +280 | 0 | +9.0 | 0 | 0 | +280 | 0 | +12.0 | 6.3 Vac | | |
| | | R | +280 | 0 | +9.0 | 0 | 0 | +280 | 0 | +12.0 | 6.3 Vac | | |

### LEGEND

\* Voltage measurements made with additional 1 meg resistor connected in series with DC probe (keep resistor leads as short as possible.)
\*\* Position: T−transmit, R−receive

### NOTES

1. Volume, squelch and R.F. gain at full CW rotation.
2. Fine tuning at mid range (knob pointer aligned with dot on panel).
3. Noise limiter switch on.
4. Voltage measurements made with a VTVM (11 megohm input).
5. All voltage measurements made from socket pin to ground (chassis).
6. Operating primary voltage is 117 VAC
7. All voltages DC unless otherwise indicated.
8. Voltage may vary ±15%.

### RESISTANCE CHART

| No. | Tube | Pin 1 | Pin 2 | Pin 3 | Pin 4 | Pin 5 | Pin 6 | Pin 7 | Pin 8 | Pin 9 | Pin 10 | Pin 12 |
|---|---|---|---|---|---|---|---|---|---|---|---|---|
| V1 | 6DS4 | | 70 k | | 3.6 meg | | | | 0 | | Fil | Fil |
| V2 | 6EA8 | 73 k | 3.7 meg | 73 k | Fil | Fil | 400 k | 0 | 2.2 k | 100 k | | |
| V13 | 12BH7 | 40 k | 150 | 220 | Fil | Fil | 40 k | 110 | 220 | Fil | | |

### NOTES

1. Power switch off
2. Volume, squelch, R.F. gain and "S" meter zero at full CW rotation.
3. Fine tuning at mid range (knob pointer aligned with dot on panel).
4. Noise limiter switch on.
5. All resistance measurements made from socket pin to ground (chassis).
6. All resistance values are in ohms.
7. Resistance may vary ±15%.
8. IC = Internal connection

**Figure 6-3:** Typical voltage and resistance charts

served when measuring the voltages and resistances. For example, if the voltage or resistance is dependent on the volume control, and the control is not set to the correct position (full clockwise in the case of Fig. 6-3), the related voltage and resistance will not be as indicated, even though there is no problem in the circuit. As discussed in Chapter 1, if you regularly service the same type of CB set, duplicate all voltage and resistance data with your own test equipment, on a known good set.

### 6-2.6 Crystal Frequency Chart

Figure 1-6 is typical of the crystal frequency charts. From a service standpoint, these charts serve a dual purpose. First, they quickly identify which crystals apply to a given channel. Second, they identify the frequency for each oscillator in the synthesizer for a given channel.

For example, assume that the frequency synthesizer covered by Fig. 1-6 is being serviced, and there is a failure of channel 1. This immediately points to a failure of the 4.765 and 16.200 MHz crystals. From another standpoint, if channel 8 is selected, the 4 MHz oscillator should produce a frequency of 4.805 MHz, whereas the 16 MHz oscillator should produce 16.250 MHz. This is the type of information you should look for when studying service literature.

### 6-2.7 Injection Voltages and Detector Output Voltages

In some detailed service literature, you will find data on how much input signal is required to produce a given output from the receiver. For example, the service literature for the Pace CB set described in Sec. 3-3 shows that for a 100 $\mu$V signal at the antenna, the receiver detector output should be 4.8 V. If the input signal is increased to 1,000 $\mu$V, the output is raised to 5.1 V. If the input is reduced to 6-9 $\mu$V, the detector output drops to 2.2 V. However, do not count on obtaining this type of data from most CB service literature!

## 6-3 SAMPLE ALIGNMENT/ADJUSTMENT PROCEDURES

The following paragraphs describe alignment/adjustment procedures for typical CB sets. The sets are selected from those described in Chapter 3. Keep in mind that these specific procedures apply directly to the sets of Chapter 3. When servicing other sets, you must follow the manufacturer's service instructions exactly. Each type of CB set has its own adjustment/alignment controls, which may or may not be different from controls on other sets.

Sec. 6-3   Sample Alignment/Adjustment Procedures   295

In the absence of manufacturer's instructions, and to show you what typical CB setup procedures involve, we will describe a complete setup, as recommended by the manufacturers. Using these examples, you should be able to relate the procedures to a similar set of controls on most CB sets. Before going into the details, here are some notes that apply to tuning circuits found in many CB sets.

### 6-3.1   Physical Location of Tuning Slugs

The service literature alignment procedures of some CB sets specify the locations of the tuning slugs with respect to the mounting board or chassis. Typical tuned-circuit assemblies are shown in Figs. 6-4 and 6-5. The coils are designed so that the proper inductance value is obtained when the tuning slug is *between* the minimum and maximum inductance range of the coil, rather than at minimum or maximum. This means that circuit resonance is obtained at two physical locations of the coil (see Fig. 6-4a) because at position 1 the slug is surrounded by as many turns of the coil as at position 2.

Now assume that a secondary winding is added at the bottom of the

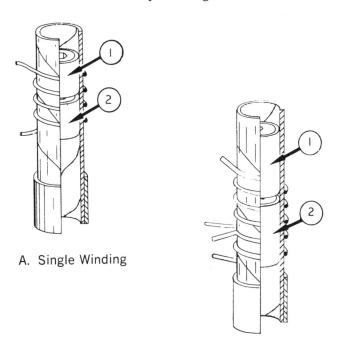

A. Single Winding

**Figure 6-4:** Core positions at resonance in tunable coil assemblies (Courtesy B & K Precision, Dynascan Corporation)

coil form, as shown in Fig. 6-4b. Resonance is still obtained at either position 1 or 2 of the tuning slug. However, at position 2, the presence of the slug in a portion of the primary and secondary of the assembly increases the coupling between the two windings, as compared to the coupling with the slug at position 1. The amount of coupling affects gain and bandwidth. Thus, to obtain the desired coil or transformer characteristics, the slug must be properly located at resonance.

The effect of slug locations is also important in double-tuned coil assemblies for similar reasons. Figure 6-5 shows a double-tuned transformer, such as a receiver section IF transformer, with the equivalent circuit in Fig. 6-5b. Either tuned circuit of Fig. 6-5a may be resonated with the slugs at position 1 or 2. The coil spacing is designed to give the desired response when the tuned circuits are resonated with the slugs in *only one position*, usually position 1. If the circuits are tuned with either slug in position 2, the coupling will be greater (possibly overcoupled). If the circuits are tuned with both slugs in position 2, greater overcoupling will occur.

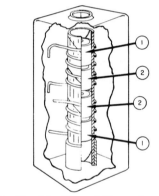

A. Double Tuned Transformer

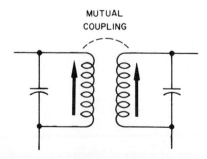

**Figure 6-5:** Effects of slug locations in double-tuned circuits (Courtesy B & K Precision, Dynascan Corporation)

### 6-3.2 Vacuum Tube Circuit Alignment/Adjustment Procedures

The following procedures apply to the Sonar FS3023 CB set described in Sec. 3-1.

*Receiver alignment.* The basic test connections for receiver alignment are shown in Fig. 6-6. The procedure is as follows:

 1. Connect an audio output meter to the external speaker jack. As an alternate, connect a VTVM or electronic voltmeter to the AVC line of $V_{6A}$ to measure AVC voltage. As another alternative, use the set's S-meter.
 2. Connect a signal generator to pin 2 of $V_3$. Adjust the signal generator to exactly 6 MHz, using a frequency counter, as shown in Fig. 6-6. Modulate the signal generator 30% at 1,000 Hz, or some other convenient audio frequency.
 3. Set the operating controls as follows: volume to one-half clockwise (about midscale), squelch full counterclockwise (full off), RF gain full clockwise (full on), fine tuning (midscale), channel switch to channel 10.
 4. Using just enough signal from the generator to give a reading of audio (or negative AVC, or an S-meter indication), align $T_{2A}$, $T_{2B}$, $T_3$, and $T_4$ for maximum indication. Always reduce signal input to that minimum which will give an indication.
 5. Inject the 6 MHz signal at pin 9 of $V_{2A}$, as shown in Fig. 6-7. Adjust $T_1$ for a maximum indication on the meters (audio, AVC or S-meter).
 6. Connect the signal generator to the antenna input, as shown in Fig. 6-8, and adjust the generator frequency to channel 10 (27.075 MHz), using the frequency counter. The meter output of the receiver should be maximum when the generator is set to exactly 27.075 MHz. If not, it may be necessary to repeat step 4.
 7. Adjust $L_1$ and $L_2$ for further maximum indication.
 8. *Neutralization of* $V_1$. Unsolder the B+ end of $R_4$ as shown in Fig. 6-9. Increase the signal generator output for AVC indication. Adjust $C_4$ for *minimum* AVC. Readjust $L_1$ and $L_2$ for *maximum* AVC. Resolder B+ end of $R_4$. Decrease signal generator output to about 1 $\mu$V. Readjust $L_1$ and $L_2$ as necessary for maximum indication.

*Transmitter alignment.* The basic test connections for transmitter alignment are shown in Fig. 6-10. The procedure is as follows:

 1. Before any attempt is made to align the transmitter, it must be assumed that no drastic adjustments have been made from the original factory settings. This same assumption applies to the receiver alignment but is of more importance in transmitter alignment.
 2. With the signal generator set for channel 10, using a frequency counter, connect the generator output to pin 2 of $V_{9A}$, as shown in Fig. 6-10. Connect a VOM, VTVM, or electronic voltmeter to the junction of $R_{52}$ and $R_{53}$ using the 3–5 V scale. The signal generator output must be between 0.5 and 1.5 V.

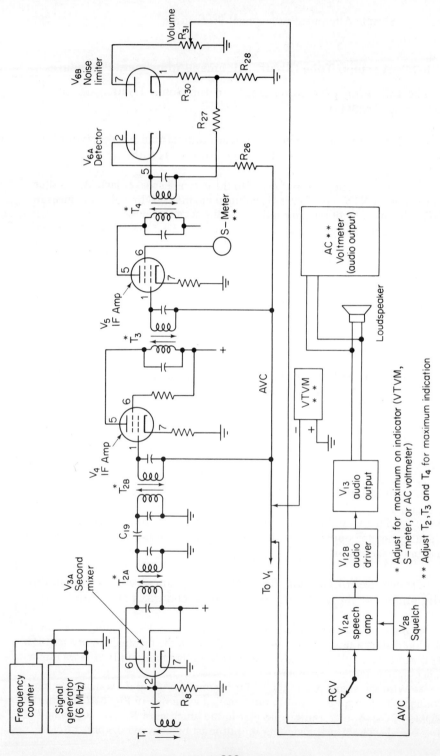

**Figure 6-6:** Vacuum tube CB test connections for IF alignment

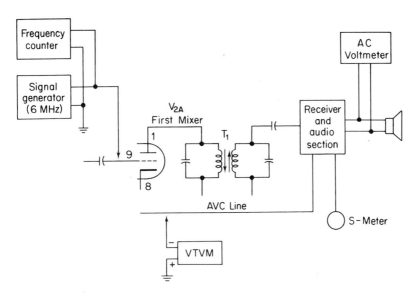

**Figure 6-7:** Vacuum tube CB test connections for first mixer alignment

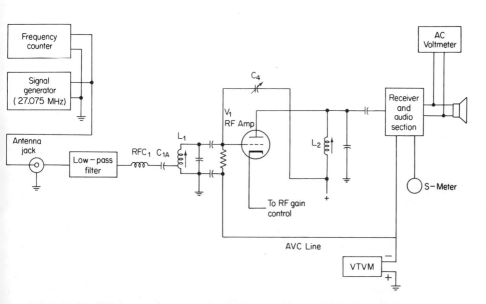

**Figure 6-8:** Vacuum tube CB test connections for RF amplifier alignment

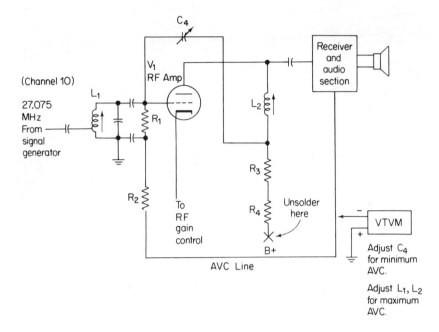

**Figure 6-9:** Vacuum tube CB test connections for RF amplifier neutralization.

3. Adjust $T_6$ and $L_4$ for maximum indication on the meter (approximately $-2$ V).

4. Move to the meter to pin 2 of $V_{8A}$. Adjust $L_3$ for maximum negative voltage on the meter.

5. Move the meter to terminal F of $T_5$. Adjust $T_5$ for maximum negative voltage on the meter.

6. Adjust plate tuning control $C_{62}$ and antenna loading control $C_{63}$ for maximum indication of front panel RF meter.

7. As a final check of transmitter alignment, measure the $V_{11}$ power amplifier grid current at the junction of $R_{52}$ and $R_{53}$, as in step 2, and disable each of the three transmitter oscillators $V_7$, $V_{8B}$, and $V_{9B}$ in turn by shorting their grids to ground. The grid current, and the front panel RF meter indication, should drop to almost zero when the oscillators are disabled if the transmitter is properly aligned. If the grid current or power output indications drop to only half, it is an indication that the transmitter is improperly aligned.

8. The trimmers of the three transmitter oscillators ($C_{28}$, $C_{38A}$, and $C_{47A}$) *should not* be adjusted unless a very accurate means is available to check the oscillator frequencies. A frequency counter with accuracies specified in Sec. 2-6 is required. Do not adjust any of the oscillators until all 23 channels have been checked. If any channel exceeds a tolerance of $\pm 750$ Hz, check the frequencies of the individual oscillators. The frequencies of $V_7$ and $V_{8B}$ are given in Fig. 1-6. The frequency of $V_{9B}$ is 6 MHz. In the case of $V_7$ and $V_{8B}$, if the trimmers are adjusted for any crystal, the frequencies must be checked for all remaining crystals; it may

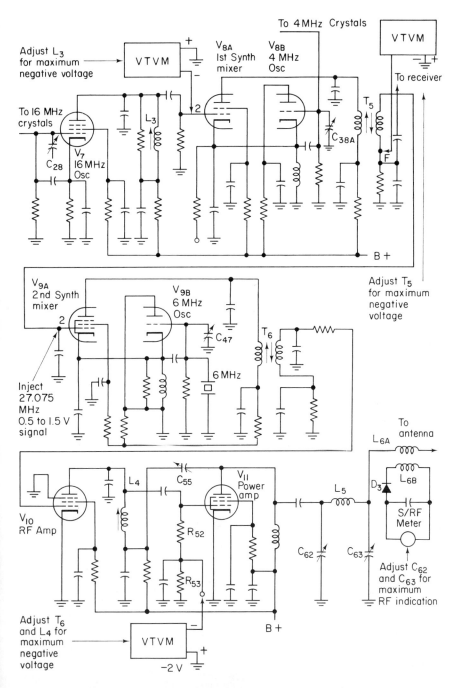

**Figure 6-10:** Vacuum tube CB test connections for transmitter alignment

be necessary to compromise. Also, if any trimmer is adjusted, recheck alignment of the entire transmitter section, as described in steps 1 through 7.

9. *Neutralization of $V_{11}$*. All of the tubes may be replaced without readjustment of the associated circuits except for the RF power amplifier $V_{11}$. If $V_{11}$ is replaced, reneutralization may be required. Neutralization is necessary when $V_{11}$ will oscillate with no grid drive applied and the antenna disconnected.

If neutralization is required, disconnect the yellow lead of modulation transformer $T_8$ as shown in Fig. 6-11. This removes B+ from $V_{11}$. Disconnect the antenna. A dummy load need not be connected, since $V_{11}$ is disabled. Connect a meter at the junction of $R_{52}$ and $R_{53}$, as in step 2, to measure about $-2$ V. Connect a meter or scope with an RF probe at the antenna. Key the transmitter and adjust $C_{55}$ for a *minimum* indication on the meter or scope connected to the antenna.

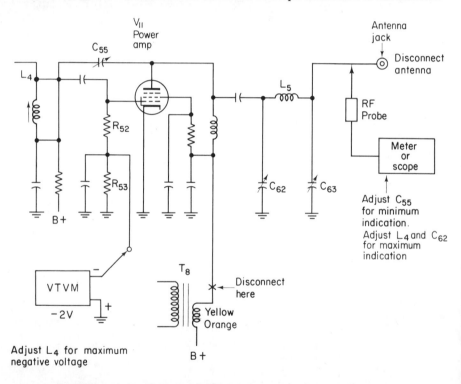

**Figure 6-11:** Vacuum tube CB test connections for neutralization of power amplifier $V_{11}$

Then adjust $L_4$ and $C_{62}$ for maximum indication on the meter or scope. Continue adjustment of $C_{55}$ (for minimum) and $L_4$–$C_{62}$ (for maximum) until no further change is noted.

10. *Measurement of $V_{11}$ plate current*. Insert a 0–50 milliammeter between the yellow lead of $T_8$ and the terminal to which it was connected, as shown in Fig. 6-12. This measures both plate and screen current. To compute plate power input, first subtract the screen current from the reading on the milliammeter.

Sec. 6-3  Sample Alignment/Adjustment Procedures    303

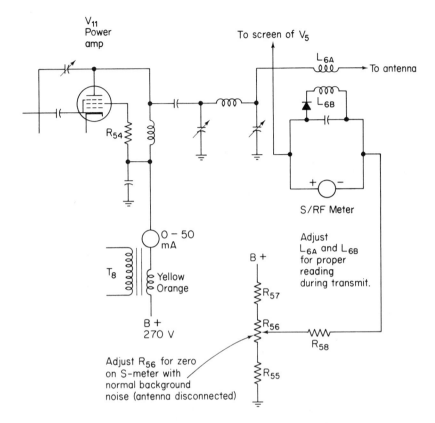

**Figure 6-12:** Vacuum tube CB test connections for power amplifier $V_{11}$ power input measurement, and S/RF meter adjustment

Screen current is computed by dividing the voltage drop across the screen resistor by the value of the screen resistor. Typically, the screen current is about 2 mA. Note that the screen resistor $R_{54}$ is selected at the factory for a power input of 5 W maximum (to meet FCC requirements). Power input is computed when the plate voltage of $V_{11}$ is multiplied by the plate current.

For example, assume that the current reading is 20.5 mA and the plate voltage of $V_{11}$ is +270 V. Allowing 2 mA for the screen current, the plate current is 18.5 mA, and the power applied to $V_{11}$ is 4.995 W (270 V×0.0185 A).

11. In the event that the relative power meter reading is too high or low, it can be adjusted by moving link $L_{6B}$ in relation to $L_{6A}$, as shown in Fig. 6-12. A higher reading is obtained when the coils are closer. Keep in mind that the meter reads *relative power only*. The true output power must be measured with an RF wattmeter (Sec. 2-8) connected to the antenna jack. Power input, as measured in step 10, must not exceed 5 W, whereas power output must not exceed 4 W. Once the coils $L_{6A}$ and $L_{6B}$ are properly adjusted, the linkage can be secured with cement.

*Modulation checks.* The RF signal output of the set is checked for proper modulation as described in Sec. 2-3. With normal voice signals directly into the microphone, the modulation percentage should be greater than 85%, but never greater than 100%. If modulation is not within these limits, adjust $R_{105}$ (in the audio clipper circuit on Fig. 3-5) to bring modulation within limits. Check modulation on several channels. The modulation percentage should remain the same.

*S-meter adjustment.* The S-meter does not normally require adjustment. As a quick check, disconnect the antenna and check that the S-meter reads *approximately* zero with normal background noise. If not, adjust $R_{56}$ (shown in Fig. 6-12) as necessary. Note that resistor $R_{58}$ is selected at the factory so that the S-meter may be zeroed by $R_{56}$ (with normal background noise). If $R_{56}$ must be adjusted to either extreme to zero the S-meter, it may be necessary to change the value of $R_{58}$.

### 6-3.3 Solid-state Circuit Alignment/Adjustment Procedures

The following procedures apply to the Pace CB144 set described in Sec. 3-2.

*Preliminary setup.* The basic test connections are shown in Fig. 6-13. Before starting alignment, set the front panel controls as follows: squelch full off, RF gain full on, delta tune to center position, CB/PA switch to CB, noise blanker off, ANL off, and volume control as convenient.

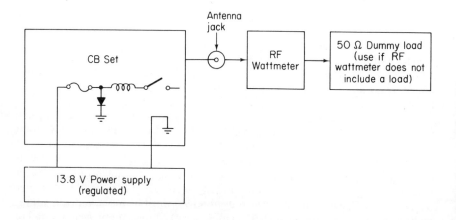

**Figure 6-13:** Solid-state CB basic test connections

## Sec. 6-3  Sample Alignment/Adjustment Procedures

$Q_{16}$ *master oscillator adjustment.* All oscillators have been precisely set at the factory. They should not be readjusted unless one of the critical tuning components associated with them have been replaced or tampered with. The test connections for $Q_{16}$ oscillator adjustment are shown in Fig. 6-14.

1. Connect a meter with an RF probe to $TP_1$.
2. Set the channel selector to channel 12.
3. Adjust $T_9$ for a maximum reading on the voltmeter.
4. Check the voltage readings on channels 1 and 23. These should be within ±10% of that obtained in step. 3. If not, readjust $T_9$ slightly to achieve this condition.

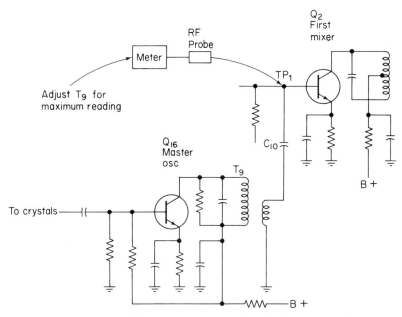

**Figure 6-14:** Solid-state CB test connections for $Q_{16}$ master oscillator adjustment

$Q_{17}$ *local oscillator adjustment.* The test connections for $Q_{17}$ oscillator adjustment are shown in Fig. 6-15.

1. Connect the frequency counter to the emitter of $Q_{17}$ through a 10 pF capacitor as shown.
2. Set the channel selector to channel 12. Rotate the front panel delta tune control $R_{80}$ to center notch position.

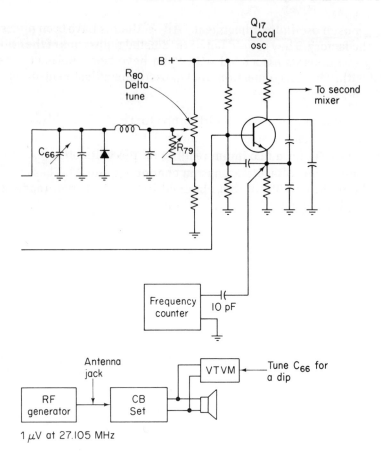

**Figure 6-15:** Solid-state CB test connections for $Q_{17}$ local oscillator adjustment

3. Adjust $R_{79}$ so that there is equal deviation (approximately ±1.5 kHz) from the center frequency when $R_{80}$ is turned to either extreme. Note that the center frequency for $Q_{17}$ is 10.595 MHz when channel 12 is selected. Thus, the frequency counter should read 10.5935 and 10.5965 MHz at the extreme positions of delta tune control $R_{80}$.

4. Return $R_{80}$ to center position.

5. Connect a VTVM or electronic meter across the speaker terminals. Connect an RF signal generator to the antenna and adjust for an output of about 1 μV at 27.105 MHz (channel 12).

6. Adjust the volume control to give a midscale reading on the meter connected to the speaker (typically about 1 or 2 V).

7. Adjust trimmer capacitor $C_{66}$ for a dip on the meter.

8. Disconnect the test equipment.

$Q_{20}$ *transmitter oscillator adjustment.* The test connections for $Q_{20}$ oscillator adjustment are shown in Fig. 6-16.

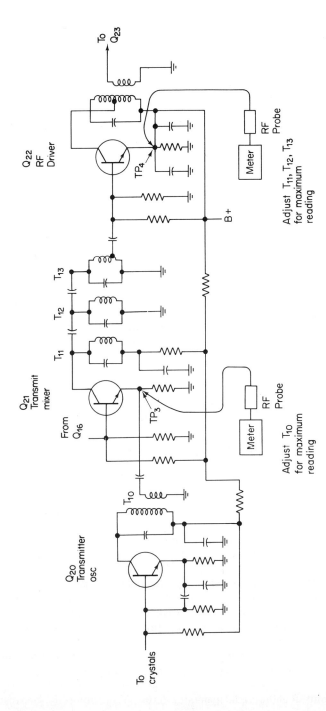

**Figure 6-16:** Solid-state CB test connections for $Q_{20}$ transmitter oscillator adjustment

307

1. Connect a meter with an RF probe to $TP_3$.
2. Set the channel selector to channel 12.
3. Key the transmitter and adjust $T_{10}$ for a maximum reading on the meter.
4. Check the voltage readings on channels 1 and 23. These should be within ±10% of that obtained in step 3. If not, readjust $T_{10}$ slightly to achieve this condition.
5. Move the meter probe to $TP_4$. Return to channel 12.
6. Key the transmitter and adjust $T_{11}$, $T_{12}$, and $T_{13}$ for a maximum reading on the voltmeter.
7. Check the voltage readings on channels 1 and 23. These should be within ±10% of that obtained in step 6. If not, readjust $T_{11}$, $T_{12}$, and $T_{13}$ slightly to achieve this condition.

*Transmitter alignment.* The test connections for transmitter alignment are shown in Fig. 6-17.

1. Set the channel selector to channel 12.
2. Key the transmitter and note that the transmit lamp lights up.
3. Adjust $L_9$ and $L_{10}$ for maximum RF output indication on the wattmeter.
4. If power output exceeds 4 W when $L_9$ and $L_{10}$ are adjusted for maximum, adjust $T_{15}$ for legal output power of 4 W.
5. With the true power output at 4 W as indicated by the RF wattmeter, adjust $R_{108}$ for a 2/3 full-scale reading on the set's S/RF meter (relative power output).
6. Check the frequency of each channel with the frequency counter. The frequency of each channel must be within ±0.005% of the channel frequencies shown in Fig. 6-1.

*Modulation checks.* The RF signal output of the set is checked for proper modulation as described in Sec. 2-3. With normal voice signals directly into the microphone, the modulation percentage should be greater than 85%, but never greater than 100%. If modulation is not within these limits, adjust $R_{92}$ (in the modulation limiter circuit, Fig. 6-18) to bring modulation within limits. Check modulation on several channels. The modulation percentage should remain the same.

*Receiver alignment.* The test connections for receiver alignment are shown in Fig. 6-19.

1. Connect the RF signal generator to the antenna jack.
2. Set the channel selector to channel 12.
3. Set the signal generator to 27.105 MHz with 30% modulation at 1 kHz.
4. Adjust the generator output for an approximate midscale indication on the S-meter (this is about an "S5").
5. Adjust $T_3$, $T_4$, $T_5$, $T_6$, and $T_7$ for maximum indication on the S-meter. Reduce the generator output as necessary to keep the S-meter at its approximate midscale.
6. Set the output of the signal generator to 100 $\mu$V and adjust $R_{72}$ for a reading of 9 on the S-meter.

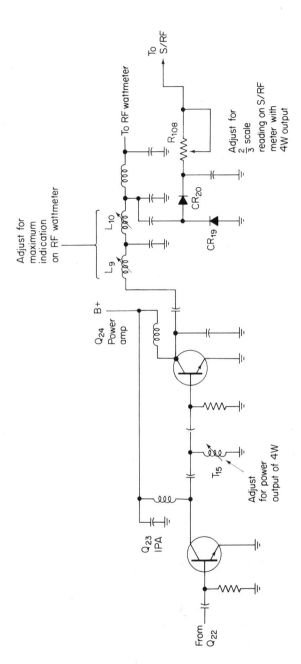

**Figure 6-17:** Solid-state CB test connections for transmitter alignment

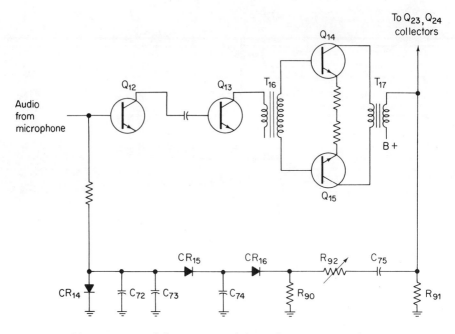

**Figure 6-18:** Solid-state CB modulation limiter circuit adjustment points

*Tight squelch sensitivity adjustment.* The test connections for squelch adjustment are shown in Fig. 6-20.

1. Set the generator to 27.105 MHz with 30% modulation at 1 kHz.
2. Set the channel selector to channel 12.
3. Adjust the generator output level to 1,000 $\mu$V.
4. Set the squelch control fully clockwise (full on).
5. Set $R_4$ to the point where less than 1,000 $\mu$V does not open the squelch.

*Noise blanker adjustment.* The test connections for noise blanker adjustment are shown in Fig. 6-21.

1. Set the generator to 25 MHz, without modulation.
2. Adjust the generator output level to 10,000 $\mu$V.
3. Connect the RF probe of the oscilloscope to the collector of $Q_9$ as shown.
4. Adjust $T_8$ for a maximum indication on the oscilloscope.
5. Change the frequency of the RF signal generator to any desired CB frequency.
6. Adjust $R_{10}$ for a minimum background noise at the speaker.

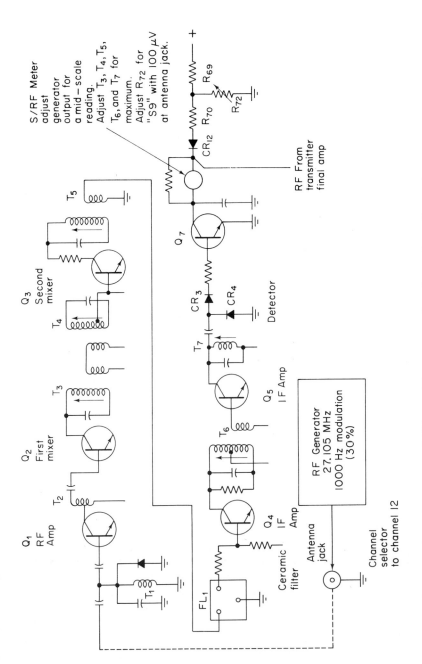

**Figure 6-19:** Solid-state CB test connections for receiver alignment

312    Ch. 6   CB SERVICE LITERATURE AND DATA

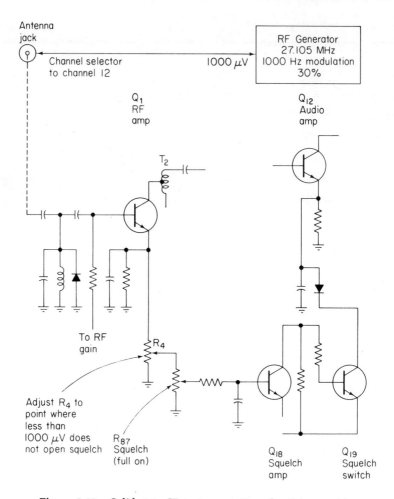

**Figure 6-20:** Solid-state CB test connections for tight squelch sensitivity adjustment

*Sensitivity check.* The test connections for sensitivity checks are shown in Fig. 6-22.

1. Set the generator to 27.105 MHz with 30% modulation at 1 kHz.
2. Set the generator output level to 0.5 $\mu$V. (Note that the receiver is rated at 0.5 $\mu$V for a 10 dB [(S+N)/N] sensitivity.)
3. Set the channel selector to channel 12.
4. Set the squelch control counterclockwise (full off).
5. Connect an ac voltmeter (with dB scale) across the loudspeaker terminals.
6. Set the volume control for a convenient reading on the dB scale of the voltmeter.

Sec. 6-3 Sample Alignment/Adjustment Procedures 313

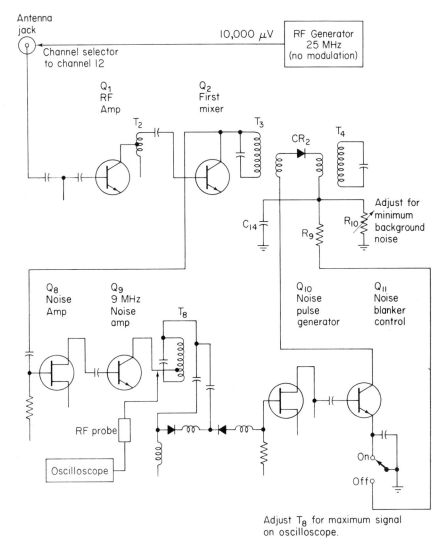

**Figure 6-21:** Solid-state CB test connections for noise blanker adjustments

7. Turn the signal generator modulation off. Do not disturb the RF output. The output indication on the meter across the loudspeaker should drop at least 10 dB. If the drop is 10 dB or greater, then the sensitivity is 0.5 $\mu$V or better.

8. If the sensitivity is 0.5 $\mu$V or better, no further checks are required. However, if the sensitivity is not 0.5 $\mu$V, or if it is desired to find the actual sensitivity, proceed to step 9.

9. Restore the modulation and increase the signal generator output for a higher dB indication. Then remove the modulation and note the drop in dB

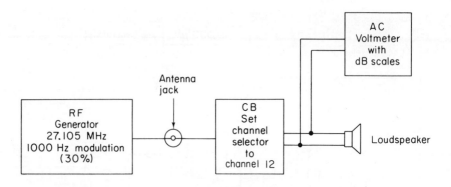

**Figure 6-22:** Solid-state CB test connections for sensitivity, selectivity, adjacent channel, image, and spurious rejection

indication. Keep working between increases in signal generator output and checks without modulation, until a 10 dB drop is obtained when the modulation is removed. Note the signal generator output amplitude at this point is equal to the receiver sensitivity.

*Selectivity measurement.* Although the manufacturer does not specify a selectivity measurement procedure, the outline below may be applied to any CB receiver. The test connections are the same as for sensitivity checks, Fig. 6-22.

    1. With the set and generator tuned to channel 12 (or any other convenient channel), squelch off, and modulation on, adjust the generator frequency for a peak reading on the meter connected across the speaker. The generator frequency should be at the channel center frequency. (If not, the set is off frequency! Check the alignment!) Note the generator output voltage, and set to a convenient value (say 5 $\mu$V).

    2. Adjust the volume control for a convenient reading on the dB scale (+6 dB, +10 dB, etc.).

    3. Double the signal generator output voltage (to 10 $\mu$V), but leave all other controls alone.

    4. Increase the signal generator frequency until the dB reading drops back to the reading obtained in step 2. Note the generator frequency as "frequency 1" or $f_1$.

    5. Decrease the signal generator frequency until the dB reading rises and then drops again to the reading obtained in step 2. Note this generator frequency as $f_2$.

    6. The frequency obtained in steps 4 and 5 are the 6 dB points, and the difference of $f_1-f_2$ is the 6 dB bandwidth. For example, if $f_1$ is 27.107 MHz and $f_2$ is 27.103 MHz, the 6 dB bandwidth is 4 kHz.

    7. Return the signal generator to the center frequency (peak reading on the meter), and reduce the generator output to 1 $\mu$V.

## Sec. 6-3 Sample Alignment/Adjustment Procedures 315

8. Again adjust the volume control for a convenient reading on the dB scale.

9. Increase the signal generator output by 100 times (to 100 $\mu$V).

10. Increase the signal generator frequency until the dB reading drops back to the reading obtained in step 8. Note the generator frequency as $f_3$.

11. Decrease the signal generator frequency until the dB reading rises and then again drops to the reading obtained in step 2. Note this generator frequency as $f_4$.

12. The frequencies obtained in steps 10 and 11 are the 40 dB points, and the difference of $f_3-f_4$ is the 40 dB bandwidth. Typically, the 40 dB bandwidth will be 20 kHz, or greater, for a CB receiver.

*Adjacent channel rejection.* Although the manufacturer does not specify an adjacent channel rejection measurement procedure, the following may be applied to any CB receiver. The test connections are the same as for sensitivity checks, Fig. 6-22.

1. With the set and generator tuned to channel 12, squelch off, and modulation on (30%, 1 kHz), set the generator output voltage to a convenient value (say 1 $\mu$V).

2. Adjust the volume control for a convenient reading on the dB scale.

3. Switch the generator, but not the set, to channel 13.

4. Increase the generator output voltage by 1,000 (to 1,000 $\mu$V) for a 60 dB measurement.

5. Note the voltmeter reading, if any. There should be no reading on the voltmeter connected across the loudspeaker with the set at channel 12 and the generator at channel 13. If there is a reading, it must be less than that obtained in step 2. If the reading is exactly the same as that of step 2, with the generator output voltage increased by 1,000 (60 dB), the adjacent channel rejection is 60 dB.

6. If the reading is less than in step 2, the rejection is 60 dB, plus the difference in readings. For example, if the reading in step 2 is +10 dB, and the reading with the generator on an adjacent channel (and increased 1,000) is +3, the adjacent channel rejection is 67 dB (10−3=7; 7+60=67). If the reading is greater than in step 2, the rejection is 60 dB, less the difference in readings. For example, if the reading in step 2 is 10 dB, and the reading with the generator on an adjacent channel (increased by 1,000) is +12 dB, then the adjacent channel rejection is 58 dB (12−10=2; 60−2=58).

*Image and spurious rejection.* The procedures for image and spurious rejection are essentially the same as for adjacent channel rejection, except for the generator frequencies. The following steps may be applied to any CB receiver. The test connections are the same as for sensitivity checks, Fig. 6-22.

1. With the set and generator tuned to channel 12, squelch off, and modulation on (30%, 1 kHz), set the generator output voltage to a convenient value (1 $\mu$V).

2. Adjust the volume control for a convenient reading on the dB scale.

3. Switch the generator, but not the set, to the desired test frequency. To find the image frequency in a double conversion receiver, find the first mixer frequency. Add the first mixer frequency to the first local oscillator frequency, if the local oscillator is on the high side (above channel frequency). Subtract the first mixer frequency from the local oscillator frequency if the local oscillator is on the low side (below the channel frequency). (In most CB receivers, the first local oscillator is above the channel frequency; high side.) In the CB 144, the first local oscillator frequency is 37.700 MHz for channel 12. The first mixer frequency is 10.595 MHz (37.700−27.105). The image frequency is 37.700+10.595=48.295 MHz. This image frequency should be rejected by the RF amplifier. To find the image frequency in a single conversion receiver, add the IF frequency (typically 455 kHz) to the local oscillator (for high side), or subtract the IF from the local oscillator for low side.

4. Increase the generator output voltage by 10 (for 20 dB), 31 (for 30 dB), 100 (for 40 dB), 316 (for 50 dB), or 1,000 (for 60 dB). These are typical values.

5. Note the voltmeter reading, if any. Usually, there will be no reading on the voltmeter when the generator is tuned to any frequency other than the channel frequency. If there is a reading, it must be less than the step 2 reading, as discussed for adjacent channel rejection. For example, the CB144 specifies a 50 dB minimum image rejection. Assume that the set and generator are tuned to channel 12 and the generator is adjusted for $1\mu V$ (step 1), the volume control is set for +10 dB (step 2), the generator is then tuned to 48.295 MHz (the image frequency for channel 12) as in step 3, and the generator output voltage is increased to 316 $\mu V$ (50 dB). Now assume that there is a reading of +7. This is 3 dB less than the step 2 reading of +10, so the image rejection is 53 dB, or 3 dB better than the specified 50 dB.

# INDEX

## A

Adjacent channel rejection, typical, 315
Adjustment procedures, 294
Adjustments, internal, 61
AGC (automatic gain control):
  circuits, 141, 149, 164, 182
ALC (automatic level control for SSB), 179
Alignment, receiver, 246
Alignment procedures, 294, 304
Alternator noise filter, 284
Amplifier troubleshooting, 219
ANL (automatic noise limiter), 12, 166
  troubleshooting, 245
Antenna:
  for CB service, 107
  disconnected, 193
  hazards, 66
  jack, 12
  loading, 12

Antenna (*Cont.*)
  servicing, 271–75
  switch, 111
  switching, 185
  tests, 273
Arc suppression at spark plugs and distributor, 278
Arcs, 193
Audio:
  amplifier, 139
  circuits, 128, 151, 185
  clipper, 131
  generator, 71
  troubleshooting, 219–44

## B

Background noise, 224
Balanced modulator (for SSB), 168
Base station, CB set, 107
Blanker, noise, 165
Block diagrams, 19, 291

Bonding systems (for interference), 286
Bracketing, 28

## C

Calibration against WWV, 92
Capacitance, distributed, 257
Capacitors:
  shunting, 194
  in troubleshooting, 210–14
Capacitor tester, 99
CB tester, 114
Channel:
  frequencies (CB), 289
  rejection, typical, 315
  selector, 12
Checkout, operational, 64
Choke coils, 280
Circuit:
  CB, 122
  groups, 31
  theory (CB set), 13–16
Clarifier (SSB), 170
Class-D CB, 2
Clipper, audio, 131
Coil, measuring inductance, 255
Coils, tuning, 295
Cold solder joints, 214
Component diagrams, 20
Control:
  CB set, 11
  settings, 194
Counter, frequency, 90
Crossover distortion, 140
Crystal:
  certified, 7
  frequency charts, 294
  tester, 99
Current measurements, 50

## D

Degraded performance, 25
Delta tune, 3

Demodulator probe, 85
Detector circuits, 127, 138, 181, 182
Detector voltages, 294
Diagrams, use of, 19, 36
Diode test, 50
Diode testers, 100
Diode testing, 207
Dip adapter, 106
Dip meter, 104
Directional coupler, 103
Distortion, audio/modulation, 223
Distortion, crossover, 140
Distortion, meter, 108
Distributor interference, 278
Dummy load, 95, 259
Duplicating measurements, 54

## E

Electronic counter, 91
Electronic switching, 185
Electronic voltmeter (EVM), 81
Engine interference, 276
Equipment:
  for CB service, 65
  failure, 25

## F

FCC rules, 2
Feedback amplifier troubleshooting, 226
Field strength meter, 100, 113, 273
Filter:
  capacitors, 279
  low-pass, 133
  RF, 161
Fine tuning, 12
Fixed station, 4
Frequency:
  CB, 289
  charts, crystal, 294

# INDEX

Frequency (Cont.)
   control, 3
   counter, 90
   meter, 90
   resonant, 254
   response, audio/modulation
      circuits, 220
   synthesizer circuits, 7, 131
   synthesizer troubleshooting,
      215–19

## G

Gain, audio/modulation circuits, 222
Gain, effects of transistor leakage, 229
Generator:
   audio, 71
   pulse, 72
   RF, 69
   signal, 68
Grid dip meter, 104

## H

Half-split technique, 41
Hazards, antenna, 66
Heterodyne frequency meter, 90
High voltage probe, 83

## I

IC (integrated circuit):
   equipment, isolating trouble, 35
   testing, 49
   troubleshooting, 208
   voltage regulator, 157
IF amplifier circuits, 126, 181
Ignition system interference, 279
Ignition system maintenance, 287
Image rejection, typical, 315
Indicators, CB set, 11

Inductance, measuring, 255
Injection:
   signals, 195
   voltages, 294
Inspection using the senses, 47
Interference filters, 279
Interference problems, 276
Intermittents, 194
Introduction to CB service, 1
Isolation transformer, 108

## L

Lamp, dummy load, 97
LC circuit frequency, 254
Leakage, transistor, 205, 229
License, FCC, 2
Linear amplifier, 8
Loudspeaker, disconnected, 193
Loudspeaker, external, 12
Low-capacitance probe, 82
Lower sideband (LSB), 175
Low-pass filter, 133

## M

Meters, 80
Microphone:
   checks, 263
   circuits, 130, 164
Mixer circuits, 124, 179
Mobile station, 4
Modulation:
   booster, 8
   checks with oscilloscopes,
      73–78, 308
   circuits, 128
   class-D, 3
   level measurement, 7
   limiter circuits, 175
   measurement, 262
   troubleshooting, 219–44

Modulator, balanced (SSB), 168
Modulator, CB circuits, 175

## N

Neutralization procedures, typical, 279
Neutralization, typical, 302
Noise:
   background, 224
   blanker adjustment, typical, 310
   blanker circuits, 141, 165, 182
   filter, alternator, 284
   limiter circuits, 127, 138, 151, 181
   problems, 271, 276–88

## O

Ohmmeter antenna checks, 273
Ohmmeter leakage tests, 205
Oscillator:
   circuits, 141, 155, 170, 179
   injection signals, 244
   troubleshooting, 215–19
Oscilloscope, 72
Operational checkout, 64
Operation of CB set, typical, 8
Overload, RF, 164
Overmodulation, 263

## P

Parts diagrams, 20
Parts lists, 291
Plate tuning, 12
PLL (phase-locked loop), 158
   troubleshooting, 215–19
Plug-in equipment, isolating trouble, 35
Portable station, 4

Power, class-D, 3
Power meter, 110
Power output, audio/modulation circuits, 223
Power supply circuits, 123, 157
Power supply for CB servicing, 108
Practical wiring diagrams, 19
Probes:
   circuits, 82–90
   compensation, 87
   pencil (generator), 68
   RF, 259
Protector, RF input, 164
Pulse generator, 72

## Q

Q of resonant circuit, 257

## R

Receiver:
   alignment, typical, 297, 308
   circuits, 123, 134, 149, 179, 186
   troubleshooting, 244–53
Relative field strength, 100
Relay, switching, 131
Repairing trouble, 63
Replaceable modules, 30, 47
Resistance:
   charts, 53, 293
   effects of voltage, 214
   measurements, 50, 53, 200
Resistor, dummy load, 95
Resonant circuit Q, 257
Resonant frequency of LC circuits, 254
RF amplifier circuits, 124, 179
RF (radio frequency):
   filter, 161
   gain circuits, 141

# INDEX 321

RF (radio frequency) (*Cont.*)
  gain control, 12
  generator, 69
  input protector, 164
  overload circuit, 164
  probe, 83, 89
  tests, 253
  wattmeter, 97, 259
RF/S meter circuits, 146, 185, 190, 261
Rules, FCC, 2

## S

Safety precautions, 66
Schematic diagrams, 19, 54, 291
Selectivity measurements, 314
Self-resonance of coil, 256
Sensitivity check, typical, 312
Service:
  approach, 192
  equipment for CB, 65
  license requirements for, 7
  literature, CB, 289
Servicing notes, 193
Shield systems (for interference), 283
Signal:
  comparisons, 37
  generators, 68
  injection, 195
  measurements, 51
  paths, CB set, 39
  substitution, 40
  testing, 49
  tracing, 40
  tracing with oscilloscopes, 73
  tracing with probe, 86
Sinewave analysis, 223
S-meter, 12, 154
Solder joints, cold, effects of, 214
Solid state service, 193
Solid state sets, testing, 49
Solid state voltages, 50

Spark plug interference, 278
Sparks, 193
Speaker, external, 12
Specifications, CB set, 11
Spectrum analyzer, 108
Spurious rejection, typical, 315
Squarewave analysis (audio), 224
Squelch:
  adjustment, 310
  circuits, 127, 140, 153, 166, 184
  control, 12
  troubleshooting, 245
SSB (single sideband) circuits, 168
S/RF meter circuits, 146, 185, 190, 261
Static, wheel and tire, 287
Stations, CB, 4
Switching:
  antenna, 185
  electronic, 185
  relay, 131
  transmit-receive, 143, 154, 161
SWR (standing wave ratio):
  description, 101
  meter, 110
  meter checks, 272
Synthesizer:
  circuits, 7, 131
  SSB, 174
  troubleshooting, 215–19

## T

Test connections, 195
Test set for CB service, 109
Testers:
  CB, 114
  diode, 100
  for CB service, 98–100
  transistor, 205
Testing:
  active device, 49–51
  capacitors, 210–14
  diodes, 207

Testing (Cont.)
  transistors, 201
Thermal runaway, 62
Tire static, 287
Transceiver, 2
Transformers, double-tuned, 296
Transients, 193
Transmitter:
  alignment, typical, 297, 308
  circuits, 128, 134, 147, 175, 190
  frequency measurement, 7
  troubleshooting, 253–70
Transistor:
  leakage, 229
  test, 50
  testers, 98
  testing, 201
  troubleshooting, 198
  voltages, 197
Transistorized probe, 86
Trouble:
  isolating, 20
  isolating to a circuit, 35
  localization, 19, 31
  locating, 21, 46
  symptoms, 18, 24, 27
Troubleshooting:
  amplifier, 219
  ANL circuits, 245
  audio/modulation, 219–44
  basic, 16
  capacitors, 210–14
  CB, 24
  feedback amplifier, 226
  frequency synthesizer,
    oscillator, PLL, 215–19
  IC (integrated circuits), 208
  with probe, 88
  receiver, 244–53
  squelch circuits, 245
  steps, relationship of, 22
  systematic procedure, 22
  transmitter, 253

Troubleshooting (Cont.)
  universal, 18
  with voltages, 198
Tuning, class-D, 3
Tuning, plate, 12
Tuning, slug, 295
TVM (transistorized voltmeter), 81
TV set for CB service, 108
Two-way radio test meter, 116–21
Type acceptance, 6

**U**

Undermodulation, 263
Upper sideband (USB), 179

**V**

Vacuum tube:
  CB, 8
  circuits, 122
  testers, 98
  testing, 47
VCO (voltage controlled
    oscillator), 158
Vibrator interference, 287
Voltage:
  CB, 5
  charts, 52, 293
  effects on resistance, 214
  IC, 209
  measurements, 51
  regulator circuits, 157
  RF, 254
  testing, 50
  transistor analyzing and
    measuring, 196
Volume control, 12
VOM (volt-ohmmeter), 81
VSWR (voltage standing wave
    ratio), 101

## INDEX

VTVM (vacuum tube voltmeter), 81

**W**

Walkie-talkie:
    CB, 3, 108
    circuits, 186

Wattmeter, RF, 97
Wavetraps, 280
Wheel static, 287
WWV calibration, 92

**Z**

Zero-beat frequency meter, 90